Teaching
with Technology:
Rethinking Tradition

Teaching with Technology: Rethinking Tradition

Printed and bound in the United States of America

Library of Congress Cataloging-in-Publication Data

Teaching with technology : rethinking tradition / edited by Les Lloyd.
   p. cm.
   Includes bibliographical references and index.
   ISBN 1-57387-068-4
   1. College teaching--Data processing--Case studies.
 2. Educational technology--Case studies.   3. Educational
  innovations--Case studies.   I. Lloyd, Les.
  LB2331.T429 1999
  378.1'733--dc21
                                   99-12230
                                      CIP

Publisher: Thomas H. Hogan, Sr.
Editor-in-Chief: John B. Bryans
Managing Editor: Janet M. Spavlik
Production Manager: M. Heide Dengler
Cover Design: Adam Vinick
Book Design: Patricia F. Kirkbride
Indexer: Laurie Andriot

Teaching
with Technology:
Rethinking Tradition

edited by
Les Lloyd

Information Today, Inc.

Medford, New Jersey

Table of Contents

Preface

The papers contained in this volume represent the efforts of teaching faculty to integrate technology into the classroom. Most of these faculty would not identify themselves as "technologists." It is more appropriate to say that they are using technology to deliver or reinforce the content in which they are experts.

These authors have shared their experiences in a variety of academic disciplines as well as shared administrative issues concerning the use of technology in the classroom. These issues include classroom design and faculty training as well as dealing with a campus climate of change represented by the intrusion of computers into the classroom.

More than fifteen years into the personal computer revolution, academics continue to deal with the cultural change needed to make full use of technology. The most difficult issue we deal with is reluctance—reluctance from some students to change the way they work and research, from some faculty to change the way they teach, and from some administrators to change budget patterns in order to provide for innovation even before there are quantifiable results. My hope, and that of the authors, is that the experiences described here will help the reader deal with related issues at his/her own institution.

Les Lloyd
Associate Vice President
Information Technology
Rollins College
Winter Park, Florida

CHAPTER 1

Technology and Classroom Design: A Faculty Perspective

Joel A. Cohen, Ph.D.
Director of Information Technology Services

Mark H. Castner
Assistant Director of Information Technology Services for User Services

Canisius College
Buffalo, NY

INTRODUCTION

A recent *Wall Street Journal* article (Wysocki, 1998), "Pulling the Plug: Some Firms, Let Down By Costly Computers, Opt to 'De-Engineer'," offers some lessons for classroom design. Classrooms may be getting too complicated, both for the instructors who want to use them and those who do not. Many technology classrooms are designed in ways that create a hostile environment for instruction. The designers seem to start with a premise about how to incorporate state-of-the-art technology rather than to examine underlying needs for creating a suitable teaching and learning environment.

This article will explore faculty preferences in classroom design, some shortcomings of current design practice, and recommendations for classroom design. Findings are based on visits to several institutions of higher education and research on faculty preferences at Canisius College. The College is currently in the process of renovating much of its instructional space. Canisius College is a private, Jesuit institution with Carnegie Class "Master's (comprehensive) universities and colleges I." Students are steeped in the liberal arts tradition. Faculty pride themselves on their concern for students as individuals. The authors' views are tempered by this environment.

Problems with Design of Instructional Technology Classrooms

One of the authors recently visited a newly renovated classroom building at a comparable institution of higher education. The architectural firm, a candidate for the renovation project at Canisius College, was showing the building off as an example of state-of-the-art technology design. The building had a very attractive auditorium that seated several hundred students. In front of each student position was a duplex outlet and an ethernet jack.

The auditorium's technology was being demonstrated by an A/V technician from the university. The technician was using a specially designed podium that incorporated a computer, VCR, connections to the local cable TV system, and laser disc player. In this demonstration, the A/V technician was using his own laptop instead of the computer in the podium. He unlocked the podium for access to the projector interface, and he faced a confusing maze of electronic equipment and cables. But he had rehearsed this part; he found the connection for his laptop and activated the podium. Impressively, the lights dimmed, the shades of the large windows around the auditorium descended, and a computer image was projected onto a huge screen. The screen was so large that it covered the entire writing board, but the image thrown from the projector covered only approximately 25 percent of the screen surface. Nevertheless, it would not have mattered if the screen were offset so that the instructor could use the writing board; the room was in nearly total darkness. Nobody would have been able to see the board anyway. The students could not see the instructor or each other. This was called the "note-taking mood." Perhaps "fall-asleep mood" would be more appropriate!

Someone asked the A/V technician, who likely had more extensive training for presentations than the average faculty member, how to adjust the lighting so that he could be seen and the students could be seen without washing out the image on the projector screen. He said he didn't know. The lighting engineer who had designed the room's lighting was present. He explained that the lighting in the room was computer controlled and could be preset in any of several pre-programmed lighting "moods." He then cycled through several of the moods using both the podium-top controls and an array of unlabeled lighting "mood" switches on the wall. None of them produced the correct lighting for student and faculty interaction while the projector was in use.

The institution has fewer than 2,200 students enrolled. What could be the instructional use for a room with several hundred fixed seats with ethernet access at each? There are probably few class sections on this campus larger than 25 students. Was the room designed to be a showcase for technology at the expense of good instructional design?

The next "high tech" room that the A/V technician demonstrated was a distance-education room. There was no problem in this room keeping everybody illuminated while the projector was on. The ceiling had lighting panels with parabolic lenses covering the entire surface. Similar to the angling of solar panels toward the sun, the parabolic lighting panels were angled away from the front of the room and toward the fixed seating. The A/V technician volunteered that faculty and students had complained that the lights in the room were too bright to the point of distraction, and that the architects, engineers, and contractors were working on the problem.

To be fair to the institution and the architects who were showing off their work, there were many things about these two rooms that worked well. Projection images were crisp, the sound systems were excellent, and the building infrastructure will support future technologies. Nevertheless, the rooms lacked an instructor's perspective. The rooms were too complicated to use and class interaction was difficult.

Lighting problems are very common. Projection screens will be washed out if too much light falls on them. Poor lighting in the audience will make it difficult for the instructor and class to interact. Glare is a frequent problem on computer screens. Poor location of lighting controls presents additional problems. Lighting controls may be placed only at the doorway so the instructor must ask a student for help with the lights. Or lighting controls may only be at the podium so that when an instructor leaves the room and conscientiously turns off the lights on the way out, some lighting is still left on.

Teaching podiums present another set of human factors problems. The size of the podiums can reach unreasonable proportions for some classrooms. Tall podiums can affect sight-lines to the instructor. To the extent that they interfere with non-technology-based classroom activity, they are an unwelcome detraction from room flexibility. Some commercially available technology podiums have extremely poor access to the equipment from the rear. This makes access to cables very difficult during set-up and later repair. Some commercially available technology podiums have a glass top, under which a computer monitor is placed.

Frequently, this results in glare problems. Videotape players are often placed too low in the podium for faculty to load without assuming an undignified position.

A podium filled with computer equipment may be a tempting target for theft. Often the equipment in podiums, and sometimes the classroom itself, is locked. Faculty must arrange to have access, and someone in the institution must administer this access. A classroom that requires special procedures for entry seems inconsistent with the need to make a classroom a place that invites students and faculty to learning. The starkness of many classrooms already invites a comparison to prison (Haney & Zimbardo, 1975).

A computer in the podium requires that many presenters share the computer. That in itself can lead to problems when faculty adjust the computer's settings to their preferences and do not return the computer to a known state. Also, the presenter must have access to instructional material that has been prepared elsewhere. In practical terms, this requires access either to a large amount of file server space or to large capacity portable media. If the material is placed on the network, the presentation may be affected by network bandwidth. If there is a Macintosh in the podium, then Wintel-using faculty will not want to use it and vice-versa. If both types of computer are placed in the podium, then the podium becomes even larger and more unwieldy.

Providing simple and cost-effective video source control is a challenge. Instructional technology classrooms often have multiple video sources feeding a projector. Faculty may be given a handful of remote controls for a projector, VCR, and laser disc player and then be expected to orchestrate a classroom presentation. If the room is connected to a cable TV system, the instructor may inadvertently show portions of unwanted TV broadcasts, often to the amusement and utter distraction of students. Automated controls have their own problems. Instructors may bypass them and operate the equipment directly, leaving the equipment in an unknown state to automated controllers.

One use of automated controls that defies intuition is centralized building control of all audiovisual equipment. In this design, all video recorders and laser disc players are located centrally. All media must be available at this central site. The instructor must schedule the use of classroom media so that it can be loaded by someone else at the proper time. Then using an automated control system, the instructor controls the equipment from the classroom. To support this operation, an elaborate audio-visual infrastructure, including cabling and control equipment, must be installed; there must be staff available at the central site to accept and schedule media; and faculty must be comfortable with the control system. In this design, a high value is placed on protecting relatively inexpensive equipment and media, and a great deal of money is needed to install and operate the centralized equipment. Most faculty are capable of finding a videotape in a library, taking it to class, and operating a VCR in the room without the need for an elaborate infrastructure. Moreover, faculty prefer decentralized systems (Baird, 1996).

MORE ON SIMPLICITY, DESIGN, AND ENGINEERING

Faculty should be able to do simple things simply, so the primary design goals for technology in a classroom should be simplicity and reliability. Design and use are simple when there are few, if any, options. As options or choices are added, either use of the technology or its design becomes more complex.

By itself, a VCR connected to a television is very easy to operate. A computer connected to a projector is similarly easy. But connect both the VCR and the computer to the same projector and use of the equipment becomes somewhat more complex. Many projectors require the user to select the input source manually. When the VCR is attached to the television, faculty are familiar with the method for adjusting the volume. When the VCR is attached to the projector, faculty must learn to adjust the volume through the projector or deal with a separate sound system. None of this is terribly difficult, but it is one more hurdle during the first week of class, when faculty wants to be thinking about other things.

Now suppose we wish to have two video sources, e.g. a VCR and a laser disc player, feeding the projector. Someone or something has to decide which source will be displayed. If we opt for *someone* to do it, then the faculty member has one more button, somewhere, to push. If we opt for *something* to do it, then we need an automatic video switching device. This switching device cannot read the instructor's mind so it switches based on an algorithm built into its electronics, an algorithm that may or may not be intuitive to the instructor.

Unfortunately, the number of choices that have to be made increases rapidly as the number of pieces of equipment in the classroom increases. The two most common sources of images, the VCR and the computer, use completely different types of signals (typically composite video and VGA, respectively), requiring a choice of which signal will be displayed by the projection system in the classroom. If a second composite video source is added, such as a laser disc player, a visualizer, or a hand-held video camera, then there is a second fork in the road at which another choice must be made. If we allow for multiple computer inputs to the projection system, we now have three points at which a choice must be made. Either the instructor must make these decisions or an automated, electronic switching device is needed at each point. (The other option is a media control system, discussed later.) Since we wish to keep use of the system simple and do not wish to burden the instructor with these choices, we introduce another piece of switching equipment at each decision point. This complicates the engineering, increases cost, adds to the possible points of failure, and puts the instructor at the mercy of the switching rules built into the equipment.

As if this is not complicated enough, most of these video sources also produce accompanying audio signals. Consumer grade audio amplifiers will handle follow-along video switching, a viable option when there are few signal sources. Commercial grade video switchers will handle a larger number of inputs and have follow-along audio. Careful labeling of a video switcher makes it reasonably easy to operate, but it is still one more button that the instructor has to remember to push.

Our classroom technology designs at Canisius College have been moving from the complex to the simple over the past several years. We started by offering instructors all the options we thought they may want: computer, VCR, laser disc player, visualizer, laptop input, external video input, and external audio input. To handle the complexity we started using a media control system to coordinate the various pieces. AMX and Crestron make the most common media control systems. They consist of a touch-sensitive LCD panel mounted on the instructor's desk or podium along with a control unit hidden in the bowels of the electronic control cabinet. They're sexy, they work, they're complicated, and they're expensive! You can program them yourself or have the vendor do it for you. When completely configured, one press on one control button on the LCD panel will lower the projection screen, turn on the projector, power up the computer, turn on the audio amplifier, select the correct source, adjust the volume, and even change the lighting.

The programming of a media control system is subject to logic errors just like any other programming. Correct programming does not guarantee that the results, such as lighting patterns or available options, will be what the faculty want. At Canisius College we have had several media control systems working for several years. They are reliable (attention to detail) and faculty like them (we included faculty in the design process), but they are also expensive and complex, and most importantly, most faculty don't use most of the options available.

As a result, we are moving to simpler models with fewer options and fewer components. Our newest design contains: a VCR wired directly to the projector for both sound and video, Macintosh and VGA cables for laptops connected to an autoswitcher connected directly to the projector, an external audio connection, an ethernet connection for the laptop, and a telephone for support. Instructors bring a VCR tape, a laptop, or an audio source; that's it. The instructor uses the remote control for the projector to select the video source and to adjust the volume. We estimate that this type of room will satisfy over ninety percent of faculty needs (for those with laptop computers), and a small number of more complicated technology classrooms will handle faculty who need other services.

RESULTS OF SURVEYS: STARTING WITH FACULTY

The place to start with classroom design is with the needs of students and faculty. Maslow (1954) described an heirarchy of human needs that ranged from essential safety and survival issues to those that resulted in comfort and self-realization. Similarly, faculty and students have basic needs for classroom space that are independent of technology. Students need to be comfortable in adult-style seating. There needs to be adequate writing board space available. Students and faculty must be able to hear each other without the interference of external noise, or internal noise from HVAC systems or projectors. Students and faculty need to see each other so that facial expression and other body language is apparent, because these non-verbal cues carry important components of communication (Short et al, 1976; Trevino et al, 1990).

Classroom furniture layout depends on class size and the preferred teaching style of the faculty. Faculty do not all have the same preferred teaching style, and a given faculty member may employ several different teaching styles in a single class period. Faculty place a high value on the ability of a classroom to be reconfigured. Some faculty favor table and chair seating, some prefer traditional tablet chair seating, some prefer fixed-tiered seating, and some prefer case-study seating. One style does not fit all.

The results of a survey administered to full-time faculty at Canisius College illustrate these points (see Appendix). Faculty preferred tiered seating in approximately 10 percent of their class sections. Faculty preferred to be able to arrange seating in nearly 70 percent of their class sections. In another survey (Cohen, 1997), 56 percent of the faculty who used a computer for presentation in class preferred that there be no permanently installed equipment podium. This was based on their experience with large podiums that were completely configured with computer and media equipment. The podium impaired seating configuration options. For a great majority of class sections, seating flexibility is key. Niemeyer (1997) has found that faculty tolerate small podiums that do not limit seating options.

Faculty may choose not to use any technology at all in the classroom. Even if equipment is available in the room, faculty does not use it all the time. But whether they use technology or not, technology should not get in the way of instruction. Less than 25 percent of the faculty at Canisius College who use computer-equipped classrooms use the computer equipment in the classroom more than once per week. Frequency of VCR and laser disc use is even lower. Almost no faculty wish to use a cable TV feed, and they object to seeing broadcast TV unless they ask for it (Cohen, 1997). In other words, not all rooms require all types of media.

Technology aside, faculty showed great interest in room comfort (see Appendix). Air conditioning, overhead projection, room lighting control, adequate writing board space, and soundproofing were rated as being of major benefit. In their comments, faculty consistently requested comfortable student seating with a large writing surface.

RECOMMENDATIONS FOR TECHNOLOGY CLASSROOM DESIGN

General

- A classroom with technology is still a classroom. To assure student comfort and an effective teaching environment, start with the basics for classroom design. Allen et al. (1996) is an excellent resource.

- Classroom learning space should follow the preferences of students and faculty rather than the preferences of technologists. Everything in the room must flow from good instructional design and practice.

- Each educational institution has a different culture for instruction. Take into account the different teaching styles of faculty on your campus. To find out what your faculty and students want in a teaching environment, you will have to ask.

- Visit other institutions of higher education to see how they have addressed classroom design for their culture.

- To the extent possible, keep the classroom flexible. The classroom is going to be around much longer than the current technology that is installed.

- Telecommunications infrastructure should include current and reasonable future needs. For Canisius College, this means that every classroom will have a network connection available for the instructor. In addition, many classrooms will have the ability to service up to 10 networked laptops. A few will be fully laptop ready, and a few will have networked computers supplied one per student. Laptop ready implies both network and power. Cable pathways will be in place for easy future installation of additional infrastructure.

- Technology and accompanying infrastructure should not get in the way of instruction when it is not being used.

- To the extent possible, standardize instructional technology rooms on a few models. Canisius College standardized on the following types of Instructional Technology Classrooms: a room with TV and VCR capability (Level I); Level I plus a networked computer and projection capability for the instructor (Level II); Level II plus networked computer capability for groups of students (Level III); and networked computer capability for each student (Level IV). (See also [Dickens, 1996], [Niemeyer, 1997] for other standardization schemes.)

- Technology rooms have operating costs, the most important of which is personnel. In the Canisius College scheme, it is estimated that a Level I classroom requires 5 percent of a full-time employee (FTE); a Level II classroom requires 10 percent of an FTE; a Level III classroom requires 15 percent of an FTE; and a Level IV classroom requires 25 percent of an FTE. A telephone hotline for support should be available in the classroom.

Equipment Control

- Keep the operation of the equipment simple. Instructions should be no more complicated than will fit on a sign posted in the classroom.

- Although it may not be intuitive, computerized control, poorly implemented, will make the classroom more complicated. Computerized control is expensive and may not be necessary.

Lighting

- Lighting is extremely important. The goal is to keep the screen dark, the instructor light, the writing board light, and the students light. With good engineering, it is possible to come much closer to that goal than the state of most technology classrooms suggests.

- An absolute must is front-to-back, rather than side-to-side, light zoning. Parabolic fixtures over the class area, lighting over the writing board that does not disperse to the classroom, and appropriate spotlights on the instructor and equipment are a good starting point for a lighting design (Boettcher & Morrow, 1995; Niemeyer, 1997).

- Design your lighting to reduce glare on the likely location of TV and computer monitors.

- To control ambient room lighting, windows must be able to be sufficiently darkened. Opaque shades or draperies provide adequate light control (Allen et al., 1996).

Projection and Video

- The required width of the projection screen is calculated by dividing the distance between the furthest student and the screen by four. For example, in a 32 foot room, the required screen width is 8 feet (Baird, 1996; Allen et al.,1996).

- Matte screens support viewing angles up to 90 degrees, but yield the dimmest images. Lenticular and glass-beaded screens result in brighter images, but viewing angles only up to 60 degrees (Conway, 1997).

- The required diagonal size of a TV monitor is one inch for each foot of distance between the furthest student and the monitor. For example, in a room in which the furthest student is 30 feet away, the required screen size 30".

• A writing board should be available and visible when video projection is used. Don't cover all available writing board surfaces with a projection screen.

Acoustics

• Projector fans, HVAC noise, and classroom building materials make for a challenging acoustical environment. Be aware of these hurdles when selecting equipment and collaborating on building design.

• There should be no noise "spillover" to adjacent classrooms. Be especially careful of film screening rooms in which the volume is likely to match that of a movie theater.

• Acoustics are particularly important in distance-education rooms.

• Voice amplification may be necessary in classrooms that seat over 75 (Baird, 1996).

Laptops

• Laptops should be easy to connect to a room's projection system. There should be no need to open locked cabinets to access the network or room projector. (Niemeyer 1997; Rochester Institute of Technology, 1998).

• Laptops simplify instructional delivery for instructors. Faculty develop presentations on the computer that will be used for the presentation and can customize that computer however they choose.

• Given the current state of battery technology, power should be present in locations where laptop use is anticipated.

Refine the Design

• Keep soliciting feedback from students and faculty. Use this feedback to update past efforts and to improve the design of future classrooms.

At Canisius College, we are working on a classroom with a simplified design resulting in simplified operation. The essential components are:

- an autoswitching VGA selector which automatically selects the active computer input (fixed computer, Macintosh laptop, or PC laptop).

- a composite video switcher which automatically selects the VCR input if present. (If you wish to use the laser disc player, you must turn the VCR off.)

- a commercial-grade audio amplifier in which all audio sources are always active this amplifier is controlled by an external master volume control mounted in an easy-to-reach location. (Faculty can potentially make a mistake and play more than one audio source at a time, but we expect this to be rare and easily remedied on the spot.)

Rochester Institute of Technology (R.I.T.) is using an intriguing podium design that incorporates many of the above design points. The podium is relatively small (a 40" by 22" top on a 29" by 18" base). The top of the podium has four cables that come through four small holes. The holes are smaller than the terminators, and the cables can slip back into the podium until the terminator meets the hole. The cable terminators plug into the ports on a laptop: Wintel VGA, Macintosh VGA, ethernet, and sound card. These connect to the room projector, network, and sound system. A duplex outlet is available for the laptop. The rest of the podium top is clear for a laptop and instructional materials. The second shelf of the podium is open. A VCR (play only) unit is secured to this surface. The VCR has a decoder for closed caption. A phone/TTY is provided to call for assistance. Below this shelf is a locked cabinet with access doors in both the front and rear of the podium. Faculty do not need access to this cabinet. They put their laptop down on the podium, plug in the proper cables, turn the projector on with a remote control, and teach. If faculty wish to show a videotape, they pop the tape in and press play. Using a simple remote control, faculty select the desired projector input: video or laptop. There are no computerized controls. The projection screen is in the corner. It is operated by hand, and its placement leaves a large portion of the writing board available for use. There is no special room security. This room design is simple for those who wish to use the equipment, and relatively unobtrusive for those who do not.

Dr. Daniel Niemeyer has designed and installed "Smart Classrooms" at the University of Colorado. The technology is simple to use; small signs near the equipment tell a faculty member all they need to know. A small lectern (22" x 14" top) with room volume control, power, sound jack, network jack, and projector jack is available in rooms that have projectors installed. The videotape player is in a corner cabinet near the podium, and controls for the video equipment and projector are at eye level. The classroom use front-to-back zoned lighting with parabolic lenses. Directed incandescent lighting provides good illumination of the instructor, class, and chalkboard while the projector is in use. Dr. Niemeyer has included tried and true media, such as slide and overhead projectors, into his design. He has modeled both the capital and operational costs of various levels of technology classrooms. Most importantly, his classroom designs start from an instructor's perspective (Niemeyer, 1997).

Conclusion

Classroom design can become over-engineered, yet perform poorly for the average instructor. Good design comes from good starting principles. Any classroom should "work" from the perspective of both students and faculty. It is possible to build technology classrooms that "work" both when the technology is employed and when it is not. Technology classrooms are still classrooms, and they need to be designed to maximize student and faculty interaction.

References

Allen, R. L., Bowen, J. T., Clabaugh, S., DeWitt, B. B., Francies, J., Kerstetter, J. P., & Rieck, D. A. (1996). *Classroom Design Manual* (3rd ed.), College Park: Academic Information Technology Services, University of Maryland.

Baird, J. (1996, Spring). Incorporating Technology into Classrooms at the University of Utah. *College & University Media Review*, 2, 23-43.

Boettcher, J. V. & Morrow C. T. (1995). Teaching Environments at Penn State. *In* D. M. Gayeski (Ed.), *Designing Communication and Learning Environments*. Englewood Cliffs, NJ: Educational Technology Publications.

Cohen, J. A. (1996). Results of Canisius College Faculty Survey on Classrooms, Fall 1996. Retrieved June 12, 1998, from the World Wide Web: http://www.canisius.edu/its/ pubs/fac-survey/class.htm.

Cohen, J. A. (1997). Survey of Instructional Technology Classrooms. Retrieved June 12, 1998 from the World Wide Web: http://www.canisius. edu/its/pubs/itcsurvey/itc.htm.

Conway, K. (1997). Master Classrooms: Classroom Design with Technology in Mind. Retrieved June 12, 1998 from the World Wide Web: http://www. iat.unc.edu/publications/ conway/conway1.html.

Dickens, J. L. (1996, Spring). Media Packages for Classrooms: Easy as 1, 2, 3. *College & University Media Review*, 2, 55-66.

Haney, C. & Zimbardo, P. G. (1975, June). It's Tough to Tell a High School from a Prison. *Psychology Today*, 26-30, 106.

Maslow, A. H. (1954). *Motivation and Personality*. New York: Harper.

Niemeyer, D. (1997). The Smarter College Classrooms Homepage. Retrieved June 12, 1998 from the World Wide Web: http://classrooms.com.

Rochester Institute of Technology (1998). Plug and Go Podiums. Retrieved June 12, 1998 from the World Wide Web: http://www.rit.edu/%7E613www/ etc/is/PodiumRooms.HTML.

Short, J., Williams, E. & Christie, B. (1976). *The Social Psychology of Telecommunications*. London: John Wiley.

Trevino, L. K., Daft, R. L., Lengel, R. H. (1990). Understanding managers' media Choices: A Symbolic Interactionist Perspective. In J. Fulk and C. Steinfeld (Eds.), *Organizations and Communications Technology* (pp. 71-94). Newbury Park, CA: SAGE.

Wysocki, B. (1998, April 30). Some Firms, Let Down by Costly Computers, Opt to 'De-engineer.' *The Wall Street Journal*, pp. A1, A8.

APPENDIX
RESULTS OF FACULTY SURVEY, OCTOBER 1996

Teaching Style

For this class section, indicate your *prevalent* teaching style by checking *one* of the following (√):

All class sections for all instructors have been recorded and results appear below.

Teaching Style	Results
Lecture	54.7%
Seminar	15.5%
Small Groups	13.3%
Case Study	1.5%
Laboratory	5.4%
Distance Education	.3%
Other (please describe to the right)	9.3%

Room Layout

Suppose you had the freedom to select a room design. For this class section, please indicate your preference for room layout by checking one of the following (√): (You may wish to refer to the drawings included with the questionnaire.)

All class sections for all instructors have been recorded and results appear below.

Layout	Description	Results
Class Style I	Combination Chair-Desk units that could be flexibly arranged in a non-tiered classroom.	38.6%
Class Style II	Chairs and modest size tables that could be flexibly arranged.	30.1%
Class Style III	Conference arrangement with all sitting around a table.	9.1%
Class Style IV	U-Shaped tables and chair arrangement with modest tiering.	7.1%
Computer Lab	Similar to Wehle Lab, Mac Lab I, or Mac Lab II	3.9%
Tiered, Fixed Seating	Similar to Tower Classrooms or Old Main 213	3.0%
Laboratory	Science laboratory	2.9%
Distance Learning	Little classroom use. Class business conducted through computer, video cassette, audio cassette, or other means.	.5%
Other	Please describe:	4.9%

Technology Requirements

Note the following brief descriptions of Instructional Technology Classrooms (ITCs). Then for this class section indicate (√) in which type of classroom you would prefer to teach this class section if it were available to you. (Choose one.)

All class sections for all instructors have been recorded and results appear below.

Technology Level	Description	Indicate the type of classroom you prefer for this class section
No special technology	A classroom that has an overhead projector	22.9%
ITC Level 1	Includes TV monitor, VCR, overhead projector. Optional Laser Disc player.	39.0%
ITC Level 2	Includes a networked computer, projector, VCR, laser disc, overhead projector, and the ability to plug in portable computer equipment to the projection system Similar to OM 214, OM 300, KAC G-21, HSC 220.	20.8%
ITC Level 3	In addition to the equipment listed under ITC Level 2, has 5-6 computer & multimedia stations that may be used to support 5-6 teams of collaborative learners.	5.8%
ITC Level 4	In addition to the equipment listed under ITC Level 2, has a networked computer for each student. Similar to the Wehle Lab or Mac Labs.	5.3%
Other	Please describe:	6.3%

Fixed/Portable Technology

Which do you prefer? (Choose <u>one</u>)	Results
Fixed instructional facilities such as those represented by ITC Level 1, 2, 3, or 4?	74.3%
Portable equipment (i.e. laptop and LCD panel) brought into room by instructor?	9.1%
Portable equipment brought into room by support staff?	13.4%
Other miscellaneous	3.2%

Room Resources

Choices will need to be made to complete classroom renovations within Canisius' ability to pay for them. On a scale of 1 (No benefit) to 5 (Major benefit), please indicate (√) how strongly you feel the following resources should be included in the classroom renovation project.

Resource	◄◄ No Benefit		Major Benefit ►►		
	1	2	3	4	5
Air Conditioning	4.0	8.0	17.4	22.4	48.3
Additional electrical outlets	18.9	20.0	34.7	15.8	10.5
Map Hangers	47.6	17.8	15.2	9.9	9.4
Overhead projector & screen	5.1	6.6	18.8	18.8	50.8
Better ability to darken room	7.9	9.9	23.6	28.3	30.4
Carpeting	39.3	16.8	24.1	9.4	10.5
Padded or upholstered chairs	32.6	17.6	22.8	15.5	11.4
Cable TV connection	35.4	16.7	24.0	10.9	13.0
Campus telephone	44.0	16.2	17.8	12.6	9.4
Clock	13.8	10.8	21.0	19.5	34.9
White writing boards	30.6	12.9	8.8	18.3	19.4
Chalk writing boards	10.5	6.3	18.8	21.5	42.9
Writing boards on side walls	22.0	13.1	25.1	21.5	18.3
Accommodations for portable computer and projection equipment for the instructor	9.5	9.5	24.2	24.7	32.1
Lounge areas outside of classrooms, but nearby	35.6	16.5	28.2	12.8	6.9
Soundproofing	11.7	12.2	21.4	24.0	30.6

CHAPTER 2

Importance of Strategic Planning in Technology

Pradeep (Peter) Saxena
Director of Information Technology and CIO

Roberts Wesleyan College
Rochester, NY

Industry has been using strategic planning activities for proactively managing technology renewal for years. It is the only way that they can leverage new and emerging technologies to stay ahead of the competition. In the educational arena, however, technology and its implementation have traditionally been left to a few pioneering professors or to the technical staff in the information technology (IT) area.

Today, more and more educational institutions are feeling the impact of the PC and Internet culture on the social structure of our society. With the advent of PCs and the Internet in our K-12 school systems, students are expecting much higher levels of technology than what most educational institutions are used to delivering. This is forcing smaller colleges and universities to provide technologies that were traditionally left for larger universities with significantly larger budgets. Today, every college and university wants to have an Internet presence. They want all their students to have PCs and access to the network from their dorm rooms. Five years ago, even the major universities had only partial implementations of these technologies.

Add to this equation the fact that technology is redefining itself every three to five years. The associated cost cycles of managing the change, upgrading equipment, and retraining faculty and staff could be astronomical. More and more educational institutions are beginning to realize that technological change will not happen without extensive planning of resources, funding, and implementation and that this planning needs the participation and buy-in of senior management. A strategic planning activity accomplishes just this.

THE STRATEGIC PLANNING PROCESS

A good strategic plan first clearly identifies the current state of the institution in terms of technology and resources. Next, it allows the leadership to determine their technological goals in keeping with the mission of the institution and the direction that they want to take in the next year, three years, and five years. It also allows them to see what they must achieve in these timeframes and the associated cost factors, both in terms of resources and funding. And finally, it allows the institution to develop short-term and long-term strategies for acquiring the resources and funding them to achieve their desired goals.

ADVANTAGES OF THE STRATEGIC PLANNING PROCESS

The advantage of the process for leadership is that they are able to foresee the benefits of technology, its alignment with their short-term and long-term goals, the costs associated with it, and what they have to do to make it happen. This makes them not only stakeholders in the successful implementation of the technologies but also proponents of the change to their colleagues and the faculty and staff of the institution. The advantage to the faculty is that they do not have to indulge in the long and frustrating process of convincing administrators of the importance of technology. In addition, if the process is done correctly, it will provide all faculty with an equal opportunity to learn and use the new technologies.

The advantage to an institution as a whole is that a strategic plan allows a high degree of standardization. There are significant financial and logistical advantages of enterprise-scale implementation and standardization of technology versus ad hoc implementations of multiple platforms. Standard platforms tend to save more money over the long term and support a much greater degree of information exchange compared to multiple platforms. They not only require less supporting staff but also offer greater quality of support since the staff can concentrate on developing their expertise in one set of technologies.

ROBERTS WESLEYAN COLLEGE

To illustrate the value of a good strategic plan to an institution, I will use Roberts as an example. For the fall semester of 1998, the technology levels at Roberts Wesleyan College (www.roberts.edu) are as follows:

- The College has a fully functional Internet, Intranet, and Extranet.

- Each of the 750 courses offered has a Web site that includes information about the faculty teaching the course, the course syllabus, schedules, assignments, and references.

- All the faculty have individualized Web pages describing their professional and personal experience.

- All 1,500 students have individualized Web pages.

- All students will have access to the college network and the Internet either through network jacks in dorm rooms or by dialing into the college's remote access service.

- The entire college's network infrastructure is state of the art.

- Almost all faculty are on Pentium computers, most with P200 or higher.

- All classrooms have the capability to hook up computers to the projection system and to the college network. Fifty percent of them have built-in Pentiums for use as teaching aids.

In July of 1997 not even one of these items was in place. Yet within one year, we have been able to find the funding and resources to accomplish all of the above and more. The single, most important reason we were able to achieve all this in such a short span of time is that we took a top-down approach to technology integration and started with a well-defined strategic plan.

Technology is an agent of change. By creating an almost architectural view of the change, we were able to visually demonstrate a plan that would fit the direction that the institution was taking. This enabled the president and the provost to buy into the process and become stakeholders. Also, it presented the faculty and the administrators with an opportunity to not only visualize the end point but also see the big picture. This helped increase communication and generated a healthy level of confidence and anticipation on the part of the institution.

We still had to work very hard to deliver on the products and services promised. However, working down from a strategic level, we were able to develop very sound implementation plans, have well managed parallel activities, and foresee and plan for exceptions and problems far better than conventional project planning cycles.

As a result, within one year we were able to lay down a solid infrastructure that the faculty and students can now exploit within the educational process. And by planning for and including appropriate funding in future budgets, we are ensuring that we will be able to continue to provide and support a state-of-the-art infrastructure.

One of the major advantages to faculty was that key ground-level issues like resources, appropriate hardware and supporting technologies, training, and help documentation were foreseen, identified, and resolved proactively instead of as a postscript. A readily available suite of technological tools was made available to all faculty. In addition, all faculty were provided with initial hands-on training to allow them to appreciate the ease of use, help remove their fears, and plant seeds of possible uses of the technology.

This preparation enabled faculty to first learn and interact with the technological tools and then determine how to use them within their own courses. This allows for a much larger implementation of the technologies available, which was our primary objective.

FAILURE TO PLAN
COULD PROVE VERY EXPENSIVE

In closing, technology is becoming an integral part of the education delivery process. Even though the cost of an individual PC continues to decrease, the overall cost of technology implementation is very significant and continues to rise. Institutions need to plan and manage technology very carefully within their campuses, both today and in the future. Failure to do so could prove very expensive.

CHAPTER 3

Is IT
Worth It?

Philip A. Harriman II
Director of Academic Computing

Heather Moir Fitz Gibbon
Associate Professor of Sociology

The College of Wooster
Wooster, OH

ABSTRACT

Administrators and funding agencies are legitimately asking whether our huge investment in information technology is improving higher education. Two surveys conducted at The College of Wooster gathered students' impressions regarding the impact of e-mail discussion groups and Web pages on their classes and they showed that, overall, students do see these technologies as beneficial. Other studies are also summarized. But what is the right question? Is it the technology that matters, or is the benefit in how it is used?

KEYWORDS

Assessment, Attitude surveys, World Wide Web, Listserv, e-mail discussion groups

INTRODUCTION

What impact is information technology (IT) having on higher education? How are students benefiting? Can we justify our huge investment in computers and networking? Administrators and funding agencies ask these questions more and more frequently. Those of us who work with IT in higher education have an instinctual sense that technology benefits learning—why else would we work in this field? Indeed, some are so enthusiastic about the promise of technology to revolutionize education that to ask the question at all seems heretical. But to avoid this question would be irresponsible. Can the impact of educational technologies be measured?

SUCCESS STORIES

In 1997, *Campus Watch* reported that a survey of students enrolled in foreign language classes at Ohio State University showed overwhelming support of technology-based, multimedia language classes compared to traditional approaches. About 75% of students said the materials and classroom experience were superior, and about 90% would like to see even more foreign language classes offered in the high-tech rooms (McBride 1997). Also in 1997, the *Chronicle of Higher Education* reported that a study produced by Jerald Schutte, an applied statistics professor at the California State University at Northridge, showed that students learning in a virtual classroom (using text posted online, e-mail, newsgroups, chat, and electronic homework assignments) tested 20% better than students who learned the material in a traditional classroom (McCollum 1997).

However, not every story has a happy ending. In 1990, *Academic Computing* published a controversial article by Marcia Peoples Halio entitled "Student Writing: Can the Machine Maim the Message?" in which she claimed that students who wrote on a Macintosh wrote more poorly than their counterparts who wrote on DOS PCs (Halio 1990). In this article, a writing instructor is quoted as saying, "Students write differently on the Mac—frankly, I think their writing is worse, and I don't think it is because they are essentially worse writers. There's something about the large print and the big margins on the Mac that seems to encourage a simple sentence structure and childish vocabulary." Slatin et al. refute Halio's findings, questioning not only her methodology but also her assumptions about how writing quality is defined (Slatin 1990). Slatin's response was in large part motivated because administrators were citing Halio as a reason not to choose Macintosh. In attempting to answer the question "Is IT worth it?" it is important to identify who is asking the question and for what motive.

Two recent news stories highlight the political nature of assessment: a *Chronicle* article discusses the debate at UCLA over their instructional enhancement initiative project, which produced a Web page for every class, and which some faculty claim threatens their intellectual property (Young 1998); and a *New York Times* article reports on the call for assessment from Linda Roberts, director of the office of educational technology at the Department of Education (Mendels 1998).

METHODOLOGY

The College of Wooster is a national, liberal arts undergraduate institution of 1,700 students in northeastern Ohio. In May 1996, May 1997, and January 1998, the college offered week-long Web tools workshops to a total of 40 faculty members, funded by grants from the college's Hewlett-Mellon Fund for Institutional Renewal. (For more information on Wooster's faculty Web workshops, see <http://www.wooster.edu/acs/workshop/>.) Following the workshops, dozens of courses were offered that incorporated Listserv e-mail discussion groups and Web pages. Partially in response to a requirement of the grants to submit reports on the outcomes of the workshop, the authors designed and conducted surveys in 1996 and 1997 to gather student responses to the use of these technologies in their classes. Faculty members distributed and collected surveys in class; nearly 100% of the students in the 20 classes surveyed participated, for a total of 339 responses.

The instrument gathered background information including class year, the student's rating of their computer skills, and whether the student owned a computer. For Listserv e-mail discussion groups, the survey asked students to indicate for what functions Listserv was used, how students rated the number of messages, how difficult it was, and what impact they believed it had. Likewise, for Web pages the instrument asked for what purposes the Web was used, how difficult students found it, and what impact it had. In 1997, we added several questions that asked whether the Listserv groups and the Web pages helped in the development of particular skills such as writing, critical thinking, discussion, research, and computer skills.

FINDINGS

The 339 student participants are distributed as follows:

	Number	Percent
First Year	214	63.1
Sophomore	72	21.2
Junior	29	8.6
Senior	24	7.1

The high percentage of first years is due to the large number of first year seminar sections that utilized these technologies.

SKILLS

Not surprisingly, most students rated their skills as average (49.4%). Only 6.2% saw themselves as novices, 16.9% below average, 25.4% above average, and 1.8% rated themselves as experts.

COMPUTER ACCESS

Nearly 70% of the students have computers in their rooms.

WHAT WAS THE LISTSERV/WEB USED FOR?

For 75.3% of the students, Listserv was used for faculty announcements, 57.4% said for posting students' assignments, 75% for discussion, 6.5% other. In terms of the World Wide Web, 64.6% of the students used it. Of those, 83.5% stated they used it because the syllabus was posted, 65.4% said it was used for readings, 54.8% for links.

QUANTITY OF MESSAGES

Most students (56.5%) found that the number of messages was fine. Only 8% thought that there were too many, 22.9% slightly too many, 10.4% slightly too few, and 1.2% too few.

DIFFICULTY OF TECHNOLOGY

Listserv e-mail discussion lists were not difficult for students. Over 50% found it easy, 28.2% fairly easy, and only 5.9% found it at all hard. Students found the Web easy to use, though not as easy as the Listserv; 42% found it easy, 25% fairly easy, 23.6% average, 9% hard, and .5% very hard.

IMPACT OF TECHNOLOGY

The Listserv overall had a positive impact on the classes. Only 9.3% said it detracted or had little impact on the class, 53.7% thought it added some value, and 22.4% believed it clearly added value.

The Web added value to the class as well, though to a lesser extent. Only 5.4% thought it detracted, 12.4% thought it slightly detracted from their course, and 21.8% thought it had no value. More positively, 42.6% thought it added some value, and 17.8% believed it clearly added value.

WHAT IS MOST VALUABLE?

In the second year of this project we expanded the survey to find out precisely what students found valuable about the Listserv and the Web. Specifically, we asked students to state how helpful they found the Listserv for providing discussion, receiving announcements, and developing thinking and writing skills. Overwhelmingly, students found Listserv most valuable for discussion and receiving announcements. A comparison of the percentage of students finding the Listserv helpful or very helpful shows that 66% found it helpful for discussion, 61% for announcements, 29% for developing thinking skills, and 8% for developing writing skills.

For the Web, we asked students if it was helpful for gaining access to readings, for developing research skills, for developing computer skills, and for posting the syllabus. Students responded that the Web was most helpful for research. Thus, 52.7% found it helpful for research skills, compared with 20.8% for access to readings, 40% for computer skills, and 24% for access to the syllabus.

EXPLAINING THE TRENDS

The technology was generally viewed favorably by students, but what variables best predict whether a student will find the technology valuable? The most likely explanations are the skill levels of the students, whether the students found the technology difficult, access to technology, the number of messages, the instructor, and how the technology was used.

SKILL LEVELS OF STUDENTS AND DIFFICULTY OF THE TECHNOLOGY

The self-reported skill levels of students had a minor but insignificant effect on whether or not they felt the Listserv added value to the class. For the Web, however, skills were more important. Of those rating their skills as low, 48.9% found the Web valuable, compared with 83.6% of those who rated themselves as above average or expert.

Similarly, there was a small but insignificant relationship between the perceived impact of the Listserv and how difficult students found it. Whether students found the Web valuable can be predicted by how difficult they found its use. Of those who found the Web difficult to use, 57.9% found it valuable. Of those finding the technology easy, 67.6% found it valuable.

But what best predicts whether students will find the technology difficult? The skill they bring does make a difference. Among those students who rated themselves novices or below average in skill, 67.9% found Listserv easy, compared with 87.9% of those who rated themselves as above average or expert. For the Web, of those rating their skills as low, 48.9% found the Web valuable compared with 83.6% of those who rated themselves as above average or expert.

The greater predictor of level of difficulty with technology, however, was the instructor. The percentage of students responding that the Listserv was easy or very easy ranged across the courses from 70% to 100%. The percentage of students responding that the Web was easy or very easy ranged from 35.7% to 95.5% across the courses.

ACCESS TO TECHNOLOGY

Another possible predictor of the impact of technology might be whether the students have computers in their rooms. This, however, proves to have no effect. Students without computers in their rooms are just as likely to view the Listserv and Web as valuable as those with computers in their rooms.

Access does, however, predict how difficult students find the technology. Of those with computers in their rooms, 85.0% found the technology easy to use, compared with 72.4% who did not have computers. For the Web, 71.0% of those who had computers in their rooms found using the Web easy, compared with 54.7% of those who did not.

What this suggests is that though it might be more work for students without computers to use Listserv and the Web, in the end it is worth the effort.

NUMBER OF MESSAGES

Since students often complain that one of the problems with the Listserv is that the number of messages received daily is onerous, we thought that perhaps this might be a predictor of the impact. However, it seems that how many messages there actually were was less important; rather it mattered how students perceived these messages.

If we look at the relationship between the students' perception of the number of messages and whether they found the Listserv valuable, we find that students who thought there were too many messages were less likely to find the Listserv valuable. When we look at the actual number of messages, however, we find something very different. The more messages sent, the more valuable students found the list (see Table 3.1 on page 32).

There is, in fact, little relationship between reality and perception. When we cross-tabulate the perceived number of messages and the actual number, we find that 19.5% of those students in classes with fewer messages reported that there were too many, compared with 48.3% of those in classes with a medium number of messages, and 34.8% of those with a high number of messages. This indicates that it is not so much the number of messages received that is important, but how these messages are used and structured in the class.

Table 3.1

Percentage of Students Rating Listserv Valuable by Perceived Number of Messages and Actual Number of Messages*

	Low	Medium	High
Perceived	81.1	78.7	69.0
Actual	25.0	75.4	83.7

*We only have data on the actual number of messages for the second year of the 3 years of workshops.

DIFFERENCES BY COURSE

Clearly one variable that might affect how students responded to the Listserv and Web is the nature of the course and the instructor. The instructor is the greatest predictor for whether or not students found the technologies had a positive impact on the class. The percentage of students in each class that responded that the technology added some or much value to the class, ranged from a low of 32% to a high of 100%.

Similarly, the Web was experienced differently by different classes. Again, most found the Web very easy, and most found the Web valuable for the classes. The percentage of students in each class that responded that use of the Web added some or much value to the class, ranged from a low of 32% to a high of 93%.

For both the Listserv and the Web, what the technology was used for (e.g., discussion, posting messages, research papers) had no effect on whether the students found it valuable.

SUMMARY OF RESULTS

Our results can be summarized as follows:

- Overall, students found Listserv e-mail discussion groups and Web pages valuable.

- Students found Listserv e-mail discussion groups more valuable than Web pages

- Whether or not a student had his/her own computer did not affect how valuable students perceived these technologies as being

- Contrary to expectations, students in classes with the most active Listserv e-mail lists complained less about the number of messages and found the list more valuable than students in classes whose list had fewer messages

- The instructor is the most significant factor in determining whether students found these technologies valuable

A possibility for further research would be to investigate how different faculty teaching styles impact student perceptions of the value of these technologies.

OTHER STUDIES

Wills and McNaught (1996) reviewed dozens of studies on the impact of technology in education and developed the following categories of evaluation:

- Quantitative data

- Student and staff attitudes and perceptions in the use of computers for learning

- Formative evaluation for instructional design of computer-based learning packages

- The effectiveness of computer-based learning in comparison to the effectiveness of conventional teaching

- Providing evidence about whether or not computer-based learning enables students to learn effectively

- Providing information from which curriculum changes can be made

In this categorization, Wooster's study and the one from Ohio State (cited in the introduction) would fall into the second category, student and staff attitudes and perceptions. Schutte's virtual statistics classroom study and Halio's article would be in the fourth category, the effectiveness of technology-based learning in comparison to conventional teaching. It is this category of study that most administrators are interested in: Technology based learning should prove to be superior to traditional methods to justify its expense. However, is this the right question?

WHAT IS THE RIGHT QUESTION?

In order to evaluate the success of information technology in education, we need to be able to define success. This was the major criticism of Halio's study, in which she based her critique of student's writing on the Macintosh on faulty definitions of good writing. Wills and McNaught use Bloom's taxonomy of education (Bloom 1956, in Wills and McNaught 1996) to help probe this issue. Bloom describes the following six levels of learning:

- Knowledge

- Comprehension

- Application

- Analysis

- Synthesis

- Evaluation

This taxonomy is a hierarchy, leading from lower-order acquisition of knowledge through comprehension to the higher order skills of analysis, synthesis, and evaluation. In their survey, Wills and McNaught found that early uses of computer-based learning were believed to be particularly useful for the lower order areas of knowledge acquisition and comprehension. This is shown in the success of technology-based training in business, industry, and the military. But which role must technology play in a liberal arts college that stresses the higher order skills?

Roy (1997) presents a different organizational scheme for the modes of teaching:
Declarative mode

- Traditional, hierarchical

- Assumes that information equals knowledge

- Symbolized by a pyramid

Interrogative mode

- Focus on response, not delivery
- Symbolized by a circle

Experiential mode

- Typified by labs, service learning
- Reflection required to make it most meaningful
- Symbolized by a spiral

Roy says that all modes have their place, and that many situations require multiple modes. In this model, Wooster's use of Listserv e-mail discussion groups would belong in the interrogative mode. The results of our survey show that this is what students find most valuable in the technology. While it may be efficient to distribute information via the Listserv or the Web, students find the highest value in discussion and the development of ideas. Both Bloom's and Roy's taxonomies are useful in describing different levels of learning; it is important to consider the intended level when evaluating the appropriateness or effectiveness of technology to teaching.

Ehrmann believes that much of the research attempting to evaluate information technology in education is misguided because of confusion about the goals of education, noting that it takes as much effort to answer a useless question as a useful one (Ehrmann 1995). There has been so much hype about technology and its potential to revolutionize education that it has been seen as a panacea; however, Ehrmann stresses that the medium is not the message. Technology in and of itself is neutral; what matters is how it is applied. Ehrmann further states that effective use of technology does not necessarily require costly specialized multimedia courseware. He says that worldware, i.e. standard applications like word processing and e-mail, can be very effective technologies in higher education. (Wooster's use of Listserv e-mail discussion groups is an example.) Ehrmann is director of the flashlight project, sponsored by the American Association for Higher Education, which is acting as a clearinghouse for information regarding evaluation of technology in higher education. For more information, consult the flashlight project Web page at http://www.aahe.org/technology/elephant.htm.

CONCLUSION

Internal surveys, such as the ones performed at Wooster, can be helpful in understanding faculty differences and student perceptions of the impact of information technology. It is important, however, to recognize their limitations as simply reporting student attitudes and that comparative studies are much more complicated. Our survey demonstrates that what is most important is not the technology itself or the native skills of the students, but rather that instructors are intentional in their use of the technology, and that they convey this intention clearly to the students.

References

Ehrmann S. C. (1995), "Asking the Right Questions: What Does Research Tell Us about Technology and Higher Learning?" *Change*, March/April, pp. 20-27.

Halio M. (1990), "Student Writing: Can the Machine Maim the Message?" *Academic Computing*, January, pp. 16-19, 45.

McBride K. (1997), "Ohio State University: Technology in Foreign Language Learning," *Campus Watch*, May 1, 1997. URL: http://www.cause.org/pub/cw/arc/cw01-may-97.txt.

McCollum K. (1997), "In Test, Students Taught On-Line Outdo Those Taught in Class," *Chronicle of Higher Education*, February 21, 1997, p. A23.

Mendels P. "U.S. Official Calls for Studies of Technology in Classrooms," *The New York Times*, April 27, 1998. URL:http://search.nytimes.com/search/daily/bin/fastweb?getdoc+ site+ site+33894+0+wAAA+%22linda%7Eroberts%22.

Roy L. (1997), "The Virtual Promised Land: Can Technology (and Budget Cuts) Transform and Redeem Us?" Keynote address at CAUSE '97, December 4, 1997, Orlando, Florida.

Slatin J. et al. (1990), "Computer Teachers Respond to Halio" *Computers and Composition*, vol. 7, pp. 73-79.

Wills S., McNaught, C. (1996), "Evaluation of Computer-Based Learning in Higher Education, "*Journal of Computing in Higher Education*, vol. 7, no. 2, pp. 106-128.

Young J. (1998), "A Year of Web Pages for Every Course: UCLA Debated Their Value," *Chronicle of Higher Education*, May 15, 1998, p. A29.

CHAPTER 4

Effects of Virtual Reality Support as Compared to Video Support in a High School World Geography Class

Yukiko Inoue
Asst. Prof. of Educational Research

University of Guam
Mangilao, Guam

ABSTRACT

Virtual reality (VR) is a new computational paradigm that redefines the interface between human and computer. VR may result in a significant improvement over traditional instruction because it is not only an interactive multimedia tool but also a learning environment that is extremely close to reality. Yet there have been few empirical studies on the use of VR as compared to that of other computerized or non-computerized educational tools. It is thus necessary to examine VR both in different scenarios and for different applications in learning and teaching.

The evaluation plan reported here addresses one aspect of such an assessment — specifically, the effect of VR support as compared to that of video support in tenth graders' learning of world geography. The reason world geography was selected was that students could "travel" to any place in the world via VR. Videos are popular in world geography classes but may not be as effective as VR because they are passive and offer only two-dimensional graphics. One world geography course (class N = 36) selected for this experiment will be composed of five units. A rotation of the treatment is going to be used: Each group is using VR for two units and videos for two units; in addition, one unit is optional for every student. Three procedures will be used for data collection and analysis: (1) To determine the effect of VR support, competency tests will be administered; (2) To determine student attitudinal responses toward VR, students will be required to complete a survey and be interviewed; and (3) To determine the tendency of returning to VR, students will be given the option of attending the VR lab, where attendance is taken.

This investigation is going to be very important but will be limited in several respects (e.g., data come from a sample of students in a single class who may not be representative of the population). As with any other technological innovations that are used for teaching purposes, VR needs to be accepted by teachers before it can be used productively. Thus, the results of this evaluation will be valuable in expediting the design and the implementation of the VR support in school curricula. Although this evaluation focuses on high school world geography, it can be expanded to such university courses as archeology, biology, and zoology.

INTRODUCTION

We live in a physical world whose properties we have come to know well. We sense an involvement with this physical world which gives us the ability to predict...where objects will fall, how well-known shapes look from other angles, and how much force is required to push objects against friction. ... A display connected to a digital computer gives us a chance to gain familiarity with concepts that are not realizable in the physical world. It is a looking glass into a mathematical wonderland (Southerland, 1965).

GENERAL DESCRIPTION OF THE AREA OF CONCERN

"Virtual reality (VR) is coming, make of it what you may" (Bricken, 1990). VR may have a significant impact on the educational systems of tomorrow. Current educational systems have been designed for an era in which human minds, textbooks, and pencils were the major tools used to store and process information. That is no longer the case. Upon entering the 21st century, education must become responsive to changing social needs and become more effective in the learning and teaching process.

Generally, VR refers to a new computational paradigm, which fundamentally redefines the interface between human and computer. In Glenn's words, "VR refers to an environment or a technology that provides artificially generated sensory cues sufficient to engender in the user some willing suspension of disbelief" (Glenn 1992). In the VR world, people believe that what they are doing is real, even though it is an artificially simulated phenomenon. Sophisticated VR can simulate sight, sound, and touch and combine these senses with computer-generated input to people's eyes, ears, and skin. Many people can share and interact in the same environment. Thus, VR is a powerful medium for learning and training.

PURPOSE OF THE PROPOSED EVALUATION

Currently, most of the media attention on VR lies in the area of entertainment (such as VR arcade machines), yet possibilities of VR seem almost limitless. Many educators, researchers, industry trainers, and software vendors predict that the use of VR for supporting school subjects may result in a significant improvement over traditional instruction, providing unparalleled and unprecedented opportunities because VR is not only an interactive multimedia tool but also a learning environment that is extremely close to reality. Students can actively participate in, and learn by experimenting with, all the options of the VR program. In spite of these promising predictions, there have been few empirical studies on the effect of VR as compared to that of other computerized or non-computerized educational tools. It is therefore necessary to evaluate the effect of VR both in different scenarios and for different applications in educational settings.

The evaluation plan reported here addresses one aspect of such an assessment, specifically the effect of VR support as compared to that of video support in a high school class. The reason world geography was selected for this evaluation was that students could "travel" to any place in the world via VR. Although videos (which combine text, picture, voice, and animation) are popular in world geography classes, they may not be as effective as VR because they offer only two-dimensional (2D) graphics and are passive forms of learning. In contrast, VR is a three-dimensional (3D) environment, challenging students to play an active role by experiencing for themselves the "inside" of an environment, such as a jungle, a desert, or the top of an iceberg.

AUDIENCE FOR THE EVALUATION

The issue of the effectiveness of VR as an educational tool will be of interest to educators as well as researchers, particularly those who are interested in teaching with technology. The second potential audience includes educational administrators and educational planners at the federal, state, county, city, and school levels. Success of VR may have important implications for course planning, budgets, purchasing of computers, and supporting software development. Finally, both hardware and software venders may be interested in this evaluation.

FEASIBILITY OF DOING THE PROPOSED EVALUATION

This proposed evaluation requires a VR program to enhance the learning of high school world geography. Although a well-developed VR program for world geography is not available now, it is possible to develop one by modifying existing commercial systems. For instance, Isaac Asimov's Science Adventure 2.0, Animal Adventure (Newsbytes, 1993, April 6), and WorldToolkit for Windows (Computer Select, 1994, February) can be suitable for this purpose. Software vendors may be willing to absorb the associated developmental costs. Finding researchers and appropriate schools for this evaluation should not be a problem. Interest in multimedia educational tools in general, and in VR programs in particular, is abundant. Funding may be secured from the National Science Foundation (NSF) or other sponsoring agencies. Once the program for demonstrating VR is developed, it is reasonable to assume that VR educational programs will become economically feasible. The proposed evaluation is, therefore, feasible.

REVIEW OF LITERATURE

Theory Relevant to the Evaluation

Technology is expanding human capacity and enhancing human reasoning ability as well as facilitating information processing that promotes new insight and depth of thinking (Lowenstein & Barbee, 1990). New technologies have vital roles to play in the transformation of educational systems, which are concerned with how people can learn most effectively. VR has the potential to be the most effective method for helping students learn and remember best (Taitt, 1993). This technology enables difficult tasks to become simpler when students practice in the VR world, a world in which mistakes are only temporary and the learning process is streamlined when students personally experience events.

Many theories or beliefs suggest that educators can design the best curricula for students by using VR. Ashton sees VR as helping educators break down barriers of race and gender because students are able to visit different countries and experience different cultures (Ashton 1992). Biocca compares the introduction of VR to that of television in 1941 (Biocca 1992). Nilan defines the characteristics of the cognitive space where VR is used as distinguished from the physical space (Nilan 1992). Schwier believes that, in the VR world, students and the system are mutually adaptive, which is extremely important to the enhancement of learning (Schwier 1993). Shapiro and McDonald assert that the increased sensory richness of VR may influence the unconscious cognitive mechanism so that the memory of what students experienced in the VR world will be judged as real events (Shapiro and McDonald 1992). Winn and Bricken believe that students learn best when they construct understanding for themselves and that VR does teach an active construction of the environment (Winn and Bricken 1992). Woodward emphasizes the possible contributions of VR technology to educational services for students with disabilities (Woodward 1992).

CURRENT EMPIRICAL STUDIES

To support the aforementioned theories or beliefs, it is useful to examine some empirical studies that have documented learning and training available in the VR world.

> Study 1: Bricken and Byrne (1992) conducted an experiment in a summer day camp, in which 59 students (ages10-15) used VR to construct and explore their own virtual worlds. Data collected by using videos captured what was happening during a 10 minute VR experience for seven days. Students answered opinion analyses about VR experiences. Informal observations were made for studying social behaviors and broad patterns of student responses to VR. The results showed that the students were fascinated by VR and expressed strong satisfaction with it. They spent a lot of

time and effort preparing their VR experiences and demonstrated rapid comprehension of complex concepts and skills, such as computer graphics, batch renderings, Cartesian coordinate spaces, and 3-D modeling techniques. This experiment concluded that VR would provide a significantly compelling creative environment for learning and teaching.

Study 2: The relationship between VR and the abilities of children to create, manipulate, and utilize mental images for problem-solving exercises was studied by Merickel (1992). Twenty-three elementary school students (ages 8-11) were divided into two groups: One group worked with VR and the other with a regular computer system. Four cognitive ability tests were then administered to the participants: The Differential Aptitude Test, Minnesota Paper Form Board Test, Mental Rotation Test, and Torrance Test of Creative Thinking. The results of the tests indicated that VR was a highly promising technology deserving extensive development as an instructional tool.

Study 3: Many students experience difficulties in learning algebra, chiefly because the symbol systems of algebra provide major stumbling blocks to the development of conceptual models, which are especially useful in learning algebra. Winn and Bricken (1992) created an experimental algebraic environment, in which students could learn through direct interactions with the algebraic systems. In this case, VR was used in ninth grade algebra classroom settings.

Study 4: In research by Shlechter, Bessemer, and Kolosh (1992), an interactive visual simulation involving VR was investigated for its effectiveness as a training device. VR allowed role playing by soldiers. A group of several hundred students were trained with the VR system and were compared to a group trained in concentration methods. Subjective evaluations by the instructors were conducted twice during the field exercises, using standard military evaluation questionnaires. Complex statistical tests were designed to analyze the data. Empirical support was found for the relative training effectiveness of the VR system over the traditional training methods. Of special interest was the increased pace of learning during the exercises.

OVERVIEW OF CONTENT

VR is a unique computerized technology whose features are perhaps not available in other technologies. Teachers may help keep students' interest by using the VR program to provide multi-sensory simulated environments occurring in real time. Students can see different parts of the world and feel that they are there. They can touch things and even smell with the more sophisticated VRs; experience in the VR world is almost real. In addition, VR is interactive and students can manipulate and control their learning environment. Thus, VR has the potential to revolutionize the learning process. It is indeed interesting to read the following scenarios of virtual learning that Wishnietsky (1992) has envisioned:

> Students who enter the virtual world could find themselves touring any city in the world. They could view the Washington Monument, the buildings of the Smithsonian Institution, and the Capitol in Washington, DC. Each student would decide which buildings to visit and what to explore. While one student visits the Capitol, another may be taking the elevator to the top of the Washington Monument. Each student decides a destination based on his/her interests and needs.

> Classes will also be able to travel to virtual reality destinations as a group. The teacher's and each student's head-mounted display will be connected to the same computer. If the computer modeled Paris, a French teacher would be able to direct students through the streets of Paris. The group could travel by boat down the Seine and eat lunch at a French café where students could order their meal in French. After the teacher finishes the tour, students are free to explore Paris on their own or exit VR and return to the physical world.

> These scenarios indicate the potentialities of VR to revolutionize the learning and teaching process. Proving that VR can enhance the learning of world geography may increase the chances that VR will be used in other high school subjects. The proposed evaluation can be

the first step in a series of similar evaluations to examine the effectiveness of VR for supporting school subjects.

METHOD

Evaluation Questions/Focus

The idea for this evaluation came from a study by Regian, Shebilske, and Monk (1992), who recommended VR for instructional technology for the following reasons: (1) There could be benefits to learning because students are experimentally engaged in the learning context; (2) The highly visual features of VR could capitalize on the disproportionately visual capabilities of the human brain; and (3) VR may one day prove to be an extremely cost-effective interface for simulation-based learning. In summary, the "realistic" and "active" learning of VR technology is beneficial for information processing because students are able to engage in full body-mind, kinesthetic learning. The following evaluation questions are raised:

1. How is learning enhanced by VR support as compared to video support in a high school world geography class?

2. What different effects exist between VR support and video support in a high school world geography class?

3. In what ways does students' satisfaction with VR support enhance their learning in a world geography class as compared to video support?

4. What kind of research is needed to assist instructional designers in developing effective VR learning environments?

GOALS AND OBJECTIVES

This evaluation examines the effect of VR support in tenth graders' learning of world geography. Based on the above research questions, three major evaluational goals are raised: (1) To develop or improve students' academic performances in a high school world geography class; (2) To develop or enhance students' positive attitudes toward learning world geography; and (3) To develop or improve students' involvement with the learning of world geography (its objectives are measured by the attendance records taken when using VR as an option). Table 4.1

presents goals, objectives, evaluation methods, and the Student Virtual Reality Experience Questionnaire (SVREQ) items related to each of the objectives.

Table 4.1 **A Summary of Goals, Objectives, Evaluation Methods, and the SVREQ Items**			
Goals	**Objectives**	**Evaluation Methods**	**SVREQ Items**
1. To improve students' academic performance in world geography.	1. Students when using VR will score higher on tests than will when using video.	1. SVREQ 2. Interview	
2. To enhance students' positive attitudes toward taking the world geography class.	1. Students will positively rate the multimedia interactive experience with VR.	1. SVREQ 2. Interview	#13, 16, 18, 20
	2. Students will report that learning with VR is more interesting than with video.	1. SVREQ 2. Interview	#3, 4, 14, 15
	3. Students will report that it is easier to learn with VR rather than with video.	1. SVREQ 2. Interview	#1, 6, 8
	4. Students will report that they enjoy learning with VR.	1. SVREQ 2. Interview	#5, 12
	5. Students will report that VR motivates their learning.	1. SVREQ 2. Interview	#2, 19
	6. Students will report that using VR is an effective way to learn.	1. SVREQ 2. Interview	#7, 9, 10, 11, 17
3. To improve students' voluntary participation in the VR supporting.	1. Eighty percent of students will attend the VR lab when the use of VR is optional.	1. Attendance record	

INFORMATION COLLECTION PLAN

Design of the Evaluation

The world geography course selected for this experiment will be composed of five units. In the experiment, a rotation of the treatment is going to be used; each group will be using VR for two units and videotapes for two units. In addition, one unit will be optional for every student. The design of the unit assignment is illustrated below.

SAMPLING PROCEDURE

This evaluation concerns the comparison of two classes (e.g., regular and gifted); however, in order to keep the initial experiment simple, only one class is examined. The one world geography class (N = 36) is randomly divided into two groups: Group A and Group B. Students randomly assigned to Group A are exposed to the VR program during units 1 and 3, and students assigned to Group B to the VR program during units 2 and 4.

Unit	Group A	Group B
1	VR	Video
2	Video	VR
3	VR	Video
4	Video	VR
5	Optional	Optional

PROCEDURES FOR GATHERING INFORMATION

Three procedures are used. First, to determine the effect of VR support, four tests are administered to the students: Two are administered following units when VR is used and two are administered following units when video is used. The testing instrument could be a kind of standardized test (preferably one that is given statewide), and a closed-ended, multiple-choice format designed specifically to correspond to the area supported by VR (but should be non-cumulative, one after each unit). Researchers should identify the test questions relevant to VR.

Second, to determine student attitudinal responses toward the VR support, students are required to complete a survey at the end of the semester and are interviewed as a follow-up to the survey. Third, to determine the tendency of returning to VR support, students are given the option of attending the VR lab and their attendance is recorded.

OVERVIEW OF EVALUATION INSTRUMENTS

The previously mentioned four competency tests will be used for Goal 1, and the SVREQ with interviews (for the entire SVREQ, see Appendix) will be used for Goal 2. Competency tests, which measure the level of world geography knowledge, are of utmost importance. An unreliable test instrument can introduce serious errors into the experiment of the evaluation (internal validity). The tests should be designed specifically to correspond to the areas supported by the VR program. The best approach is to look for an existing test that has been validated; in such a case, there is no need to pilot test this instrument. If a special instrument is designed, however, it should be pilot tested to assure that the students understand the meanings of questions and that the questions really measure their knowledge of geography.

The SVREQ has three parts (attitudinal responses, open-ended questions, and demographic information) that total 30 items. Twenty items measure the satisfaction and usefulness of VR support; 10 items measure the perception of computers in general. Each item consists of a statement and a five-point Likert scale (1 = strongly disagree, 2 = disagree, 3 = not sure, 4 = agree, 5 = strongly agree). To avoid any response bias, items are arranged in random order. The SVREQ also includes demographic data regarding gender and ethnic backgrounds, which may impact the perception of the VR program.

VALIDITY AND RELIABILITY

Since the proposed experiment will be well controlled, it should be fairly validated. The following areas, however, need special attention:

Test Validity

As stated before, competency tests should measure the knowledge level in world geography. The best method for doing this is to use existing tests that have been validated. If special tests are designed, they should be pilot tested to assure that students understand the meaning of each question. Regarding the survey instrument (SVREQ), one effective way to determine its content validity is to use a panel of persons to judge how well the instrument has met the standards. Several judges (e.g., computer instructors, VR experts, educational consultants, measurement experts, and the teacher of the class) will examine all the items in the SVREQ. In order to have adequate content coverage, it is important for the judges not to define "content" too narrowly.

Reliability

In this evaluation, reliability can be improved if external sources of variation are minimized. Thus, the researchers can achieve enhanced equivalence through improved investigator consistency by using only well trained, supervised, and motivated persons to conduct this evaluational research. With measurement instruments such as competency and attitude, the researchers can increase equivalence by improving the internal consistency of the tests.

Pilot–Testing

The preliminarily SVREQ is pilot tested to ensure that students will respond in accordance with instruction. It will be revised and pilot tested again with personal interviews of students. After the third trial, there should be only minor adjustments for further revision in reproducing the final version of the SVREQ (for the draft of SVREQ, see Appendix).

PROTECTION OF THE PARTICIPANTS

Since participation in the experiment could have a negative impact regarding the grade achieved in the class (thereby affecting some students negatively), precautions should be made in order to protect the participants against any psychological or other harms. It is advisable to obtain the consent of the students and possibly their parents.

OVERVIEW OF DATA ANALYSIS AND INTERPRETATION

Scores are determined for each student using the results of competency tests when the student is supported by VR and when the student is supported by video. A T-test is performed to determine if significant differences exist in academic performance between the VR support and the video support. Analysis of variance (ANOVA) is performed to determine group differences and interactive effects. Additional tests will be conducted to determine if any significant difference exists due to student gender, ethnicity, and computer experience.

Descriptive statistics indicating means and standard deviations will be generated by the results of the survey instrument to assess the student attitudinal responses toward the VR support. The results of student comments and suggestions in interviews are used to determine if the results of the survey are reflecting actual compliments and criticisms. Finally, VR lab attendance records for those students participating in this experiment are analyzed. The data will be used to determine the percentage of those students that elect to continue using VR when the use is optional, and the frequency with which they would use it.

CONCLUSION

Virtual reality could become the most important computerized, multimedia technique in educational systems of the 21st century. Thus, this investigation is going to be very important, but it will be limited in several respects. First, it will only be based on data from a relatively small sample of students in a single class who may not be representative of the population. Second, there is no randomization because every student in the class participates in this experiment. Randomization occurs only when dividing into groups A and B. Third, VR support is given at different times; if sequence is not important, and if there is enough VR equipment, it is possible to use VR with both groups at the same time. Fourth, only one teacher will be involved. If the teacher is biased for (or against) VR support, it may influence the results of the evaluation.

In conclusion, and as Taitt (1993) maintains, VR has the potential to be the most effective technology or environment for helping students accelerate their learning and retention of information. It usually takes 10 years for a new technology to be widely accepted, but it is necessary to prepare for the time when VR is readily available as a learning tool. More importantly, as when using any other technological innovation for teaching purposes, VR needs to be accepted by teachers before it can be used productively in educational systems. The results of this evaluation, therefore, will be valuable in expediting the design and the implementation of VR support in high school curricula.

Although this evaluation plan focuses on a high school subject, world geography, it can be expanded to such university courses as archeology, biology, and zoology. After all, 21st-century students must master sophisticated information-age learning media, having access to more powerful learning resources than students of today.

References

Ashton K. (1992). Technology and interactive multimedia: Identifying emerging issues and trends for special education. Washington, DC: COSMOS Corporation. (ERIC Document Reproduction Service No. 350 767).

Biocca F. (1992). Communication within virtual reality: Creating a space for research. *Journal of Communication*, 42(4), 5-22.

Bricken W. (1990). *Learning in Virtual Reality.* Washington University, Seattle: Washington Technology Center. (ERIC Document Reproduction Service No. 359 950).

Glenn S. (1992, September). "Real fun, virtually: virtual experience amusement and products in public space entertainment." In S.K. Helsel's *Proceedings of Virtual Reality*, Westport: Meckler, pp. 62-63.

Lowenstein R., Barbee D. E. (1990). *The new technology: Agent of transformation.* In D.E. Barbee & G.D. Ofiesh (Eds.). A reprint on the Nationwide State of the Art of Instructional Technology. Department of Labor, Washington, DC: Office of Strategic Planning and Policy Development.

Merickel M. L. (1992, October). A study of the relationship between virtual reality and the ability of children to create, manipulate and utilize mental images for spatially related problem solving. Paper presented at the annual convention of the National School Boards Association, Orlando, Florida. (ERIC document Reproduction Service No. 352 942).

Nilan M. S. (1992). Cognitive space: Using virtual reality for large information resource management problems. *Journal of Communication*, 42(4), 115-135.

Regian J. W., Shebilske W. L., Monk J. M. (1992). Virtual reality: An instructional medium for visual-spatial tasks. *Journal of Communication*, 42(4), 136-149.

Rheingold H. (1991). *Virtual Reality,* New York: Summit Books.

Schwier, R. A. (1993, June). A taxonomy of interaction for instructional multimedia. Paper presented at the Annual Conference of the Association for Media and Technology in Education in Canada. Vancouver, British Columbia, Canada. (ERIC Document Reproduction Service No. 352 044).

Shapiro M. A., McDonald, D. G. (1992). I'm not a real doctor, but I play one in virtual reality: Implications of virtual reality for judgments about reality. *Journal of Communication*, 42 (4), 94-114.

Shlechter T. M., Bessemer D. W., Kolosh K. P. (1992). Computer-based simulation systems and role-playing: An effective combination for fostering conditional knowledge. *Journal of Computer-Based Instruction*, 19(4), 110-114. (ERIC Document Reproduction Service No. 457 936).

Southerland I. (1965). The ultimate display. In W.A. Kalenich's *Proceedings of the IFIP Congress*, pp. 506-508.

Taitt H. A. (1993, March). Technology in the classroom: Planning for educational change. In *Curriculum Report*, 22 (4), Reston, Virginia: National Association of Secondary School Principals. (ERIC Document Reproduction Service No. 359 922).

Winn W., Bricken W. (1992). Designing virtual worlds for use in mathematics education: The example of experiential algebra. *Educational Technology*, 32(12), 12-19.

Wishnietsky D.H.(1992). Hypermedia: The integrated learning environment. Bloomington, Indiana: Phi Delta Kappa Educational Foundation.

Wolley B. (1992). *Virtual World: A Journey in Hype and Hyper-reality*, Oxford: Blackwell.

Woodward J. (1992, June). Virtual reality and its potential use in special education: Identifying emerging issues and trends in technology for special education. Washington, D.C.: COSMOS Corporation. (ERIC Document Reproduction Service No. ED 350 766).

APPENDIX

Student Virtual Reality Experience Questionnaire

Part 1 - Perceptions of VR and video programs

Please use the following scale to rate each statement and circle the number that best describes your answers. (Note: Virtual reality is abbreviated as "VR" in the statements.)

1 Strongly Disagree (SD)
2 Disagree
3 Not Sure
4 Agree
5 Strongly Agree (SA)

Factor One:
Usefulness and Satisfaction of the VR Program (Circle One)

	SD				SA
1. I found the VR program easy to use.	1	2	3	4	5
2. The VR program motivated me to learn.	1	2	3	4	5
3. The VR program was dull and uninteresting.	1	2	3	4	5
4. I would prefer to learn from the VR program rather than from the video program.	1	2	3	4	5
5. The VR program was enjoyable and educational.	1	2	3	4	5
6. The VR program was not easy to understand.	1	2	3	4	5
7. I believe that the VR program was not effective for educational use.	1	2	3	4	5
8. The VR program was user-friendly.	1	2	3	4	5
9. I could learn faster using the VR program than using the video program.	1	2	3	4	5
10. The VR program did not help increase my understating of the world geography context.	1	2	3	4	5
11. I could not clearly understand the material presented in this VR program.	1	2	3	4	5

12. I believe that the VR program would be an
    excellent educational tool.   1 2 3 4 5

13. Three-dimensional presentations helped me
    to learn.   1 2 3 4 5

14. I would prefer to learn with a video-based
    class rather than from a VR-supported class.   1 2 3 4 5

15. I believe that I could learn more in other
    subjects if VR programs such as this one
    were available.   1 2 3 4 5

16. The simulated environment of the VR program
    enhanced its educational value.   1 2 3 4 5

17. The VR program was not an effective way to
    learn about world geography.   1 2 3 4 5

18. I would appreciate the interaction with the
    simulated world provided by the VR program.   1 2 3 4 5

19. More VR programs are needed for enhancing
    student learning in other subjects.   1 2 3 4 5

20. The pictures, graphs, and sound in the VR program
    did not help me learn the material presented.   1 2 3 4 5

Factor Two: Computer as a Learning Tool (Circle One)

	SD				SA
1. Computers are important to my future goals.	1	2	3	4	5
2. I feel at ease learning by using computers.	1	2	3	4	5
3. If given a choice, I do not want to learn from a VR type of computer program.	1	2	3	4	5
4. I feel confident in my abilities to work with computers.	1	2	3	4	5
5. I do not think that computer technologies will be useful in learning school subjects.	1	2	3	4	5
6. I would rather read a textbook than learn from a computer lesson.	1	2	3	4	5

7. I believe that the use of computers is not an
   effective method of instruction.   1 2 3 4 5

8. Interactive computers such as VR computers
   are more exciting than lectures.   1 2 3 4 5

9. I would prefer to learn in a traditional
   instructor-based class rather than in a
   computer-supported class.   1 2 3 4 5

10. The layout of the computer screens makes
    it easy to follow the content of lessons.   1 2 3 4 5

Part II - Comments and Suggestions about the VR Program

For the following questions, write your answers about the VR program you
have used in the world geography class.

1. In what ways do you think the VR program was effective as an
   educational tool?

   _____

   _____

   _____

   _____

   _____

2. Name the two major strengths and two major weaknesses of the
   VR program.
   Strengths:
       (1) _____
       (2) _____
   Weaknesses:
       (1) _____
       (2) _____

3. Are you (check one):
   ____ Male   ____ Female

4. Are you (check one):
   ____ American Indian   ____ Asian/Oriental   ____ Hispanic/Latin
   ____ White/Caucasian   ____ Black/African-American
   Other, please specify: _____

5. Do you have a computer at home? (Check one)

_____ Yes _____ No

6. How often do you use computers at home or at school? (Check one)

_____ Always _____ Frequently _____ Sometimes

_____ Seldom _____ Never

7. In what way do you use computers at home or at school? (Check all appropriate answers)

_____ Word processing _____ Drill-and-practice _____ Games

_____ Internet _____ Electronic mail

Other, please specify: _____

8. Did you know about virtual reality before taking this course? (Check one)

_____ I knew nothing about virtual reality.

_____ I had some knowledge about virtual reality.

_____ I had a lot of knowledge about virtual reality.

_____ I experienced the virtual reality environment.

9. Would you like to use VR programs in other school subjects? (Check one)

_____ Yes _____ No

10. If yes, in which school subjects would you like to use VR support? (Name high school subjects)

_____

_____

_____

Thank you very much for your cooperation!

CHAPTER 5

Changing Students, Changing Classroom Landscapes: Meeting the Challenge in the Small Liberal Arts Institution

Elizabeth Burow-Flak
Asst. Prof. of English

Doug Kocher
Chair and Assoc. Prof. of Communications

Ann Reiser
Assoc. Prof. of Education

Valparaiso University
Valparaiso, IN

The first generation of students who began high school after access to the Internet mushroomed in the mid-1990s are just entering college. As a full range of Internet applications has become available in schools and homes, more and more freshmen have posted Web pages, surfed oceans of information, and chatted in real-time and asynchronous environments before graduating from high school. Even first-year students inexperienced in computer use can catch up quickly, thanks to Internet connections in dorm rooms, as well as classrooms and labs. Such experience creating, consuming, and processing new forms of information has already brought new cultures to the classroom landscape that instructors need to address. For example, as students literate in Internet communication grow increasingly independent in charting both cognitive and hypertextual connections between bodies of knowledge, they may also resist facets of "traditional" instruction: information in long units; texts without navigational aids (such as icons and frequent paragraph breaks); and pedagogical styles that allow others to control the pace, such as lecture and the Socratic method.

How can instructors in the liberal arts—often, in environments with limits on technology and instructional support—appeal to students who are learning new forms of communication and equip them for a world that requires further development of those literacies? How can we, in the words of Brent (1997), help our students "stay afloat in the electronic sea of discourse," even as we buoy them on what they perceive as the "alien medium of print academic discourse?" How can we encourage students to learn and research independently while still preserving academic community, critical evaluation of sources, and the focus on challenging and sometimes difficult material? This chapter charts the efforts of three instructors to achieve the above goals in the fields of communication, composition, and education while navigating new information technologies with their students.

In the first section of this chapter, "Decision Points in Implementing Online Learning," Doug Kocher focuses on the delivery of content with new technology. He examines the often-overlooked decision points instructors should evaluate before offering online classes. Drawing on his experiences from teaching online for the past three years, Kocher reviews the challenges of beginning an online class and shows how course format and content can be affected by an institution's resource imperatives.

In the second section of this chapter, "Composing for the Web: Toward Interconnected Rhetorics in College Composition," Elizabeth Burow-Flak details the benefits of integrating print and electronic composition early in the college curriculum. Supported by sites that her first-year composition and literature students have constructed, Burow-Flak suggests how Web construction can result in better writing and vice versa. She additionally suggests how to integrate Web construction and effective use of electronic resources in upper division classes, with an eye toward new academic communities that those projects can forge.

Finally, in "Personalized Electronic Textbooks: Shaping a New Learning Environment for Professional Educators," Ann Reiser addresses how interacting by way of the World Wide Web allows for individualized and engaging projects that students cannot receive through standard channels such as lecture, research, and experimentation. Reiser also documents how preparing K-12 educators to be proficient in technology can provide both the impetus and the means for positive educational change in both university and K-12 schools. Together, the sections form an institutional portrait and demonstrate ways to meet the challenge of new literacies within reasonable budgetary and instructional constraints.

DECISION POINTS IN IMPLEMENTING ONLINE LEARNING

Much of the developmental work in creating technology-based course delivery rests, quite properly, on content issues. The content, after all, provides the dominant rationale for its existence. Unfortunately, many an educator has labored long only to learn that institutional capabilities of supporting technology are limited or in other ways pose impediments that compromise the instructor's intentions and goals.

In his review of journalism educators' attitudes and perceptions about the use of new media, Panici (1998) reports:

> Respondents also disagreed that their institution has sufficient funding for equipment, software, and resource materials to integrate new media into the classroom and offers time and support to adopt existing instruction and develop new instruction suitable for using new media in the classroom (p. 58).

Reasons for perceived lack of support vary among institutions, but certainly include those enumerated by Soules and Adams (1998):

> Buildings and equipment must be wired, installed, tested and readied to whatever level is required. Similar wiring and equipment, maintenance, and troubleshooting are needed for the faculty member in his or her office and also, possibly, at home (p. 51).

Soules and Adams write that, moreover:

> Regular maintenance is needed, along with intermittent troubleshooting (immediate and long-term). Do we have staff on hand during all class hours? Do we have enough to cover multiple problems in multiple classrooms at once? Can we "re-set" the equipment between classes so that the next faculty member can run a completely different program (p. 51)?

These issues and others affected the development of a course called "Communication 110—Introduction to Internet Communication" at Valparaiso University (hereafter referred to as COMM 110). What follows are decision points that I, as instructor for that course, needed to address in order to allow for successful instructional delivery at that institution.

Background

COMM 110's origins date back to the fall of 1992, when Internet access at Valparaiso University became a reality. While few faculty offices at that time had network connections, modem access to VAX shell accounts was possible. Only a few labs provided network access, and only via command line-based software on the University's VAX 4000. Shortly after general campus Internet access began, early versions of the first World Wide Web browser, Mosaic, became available to some faculty, but not on any network.

The predecessor course for COMM 110 was offered in a face-to-face format in a campus computer lab; it evolved into an online course for a first-time offering in the spring semester, 1995. By that point, campus Novell networks had become a reality, but were unstable. On the other hand, the VAX was quite stable; thus, COMM 110 was developed as a command line-based course delivered through VAX class accounts via e-mail, with file support from an ftp server in the instructor's office. WWW access was via the character-based browser, Lynx, which was available on the VAX. The class was delivered this way through the summer of 1998, when it was converted to a Netscape-based offering using a closed campus intranet. By that date, the Novell networks had become relatively stable and available. [1]

Decision Points in Course Development

Throughout its development, from 1992 to the present, COMM 110 has been heavily influenced by resource imperatives. The ones that follow by no means represent all of them, but they are presented as dominant shaping forces in COMM 110. They will undoubtedly also factor in course decisions at other institutions, particularly smaller ones that are focused on the liberal arts.

1. *"What kind of students am I trying to reach?"*

   COMM 110 sought a majority of students from off-campus, because it was conceived of as a distance-based course that could be taken from home. The course would require that students had adequate equipment and would be able to negotiate software on their own. They were so informed in writing, prior to registration, on the course information handout. For those students, the resource burdens are shifted from the institution to the students, though some students use campus computer labs to access the course.

2. *"Who will provide technical support?"*

   The instructor is strongly discouraged from doing this, even if able. A willingness to do so can escalate students' expectations and must be provided impartially to all if provided to any. A course information sheet should specify what responsibilities belong to the student. The instructor should be very aware of the campus computing staff's capabilities, and the Help Desk's competence, and know to whom questions should be referred regarding technical support.

   The instructor must take care that hardware and software problems do not become a distraction, for either the instructor or the campus computing personnel. Some minimal level of student responsibility and familiarity has to be assumed, and prospective students must both understand and agree with this. Otherwise, the instructor (or someone) will be inundated with debugging requests.

3. *"What hardware/software will be necessary to access the course?"*

   Here the instructor must weigh the extent to which low-end equipment, including slower modems, will be an impediment to learning. If both PCs and Macs are supported, will the institution's computing center personnel be prepared to provide basic assistance? Are advanced graphics capabilities necessary on the students' computers? Are sound and video required?

4. *"How much of the class will be taught online?"*

The more the class is taught online, the more pressures will be placed on campus instructional resources and on students with home computers. Thus, issues in questions 1 and 2 (above) become more serious as the class is taught more extensively online. When lab sessions prove necessary, the instructor must compete with other faculty needing similar resources.

5. *"What level of student access monitoring will be used?"*

Do institutional resources permit record-keeping (logging) of when and how often students access online course materials? This can become an issue if there is doubt about whether students have spent adequate time with online resources related to the course. Instructors should inquire whether campus Web server management and software can provide this information.

6. *"How much E-mail is the instructor willing to handle, and via what routes?"*

Many ways of online teaching will entail some level of E-mail interaction with students. What are the capabilities of the campus e-mail system? What accounts will be supported (local student accounts or those from any ISP, such as ibm.net, AOL, etc.)? The quality of instructional interaction can be diminished if the e-mail load is significant.

7. *"What campus computing facilities are required for course delivery?"*

Will there be occasions when the class must meet as a group in a lab setting, or can all delivery take place online at the individual's home computer/workstation? Is a lab setting needed for exams and other testing? Will this pose scheduling conflicts with other classes?

These are by no means all of the decision points that factor into online course delivery, but they are likely to represent common ones for instructors at many smaller institutions. Figure 5.1 presents each of these issues as resource decision points on a continuum, with X an approximation of the instructor's decisions for COMM 110.

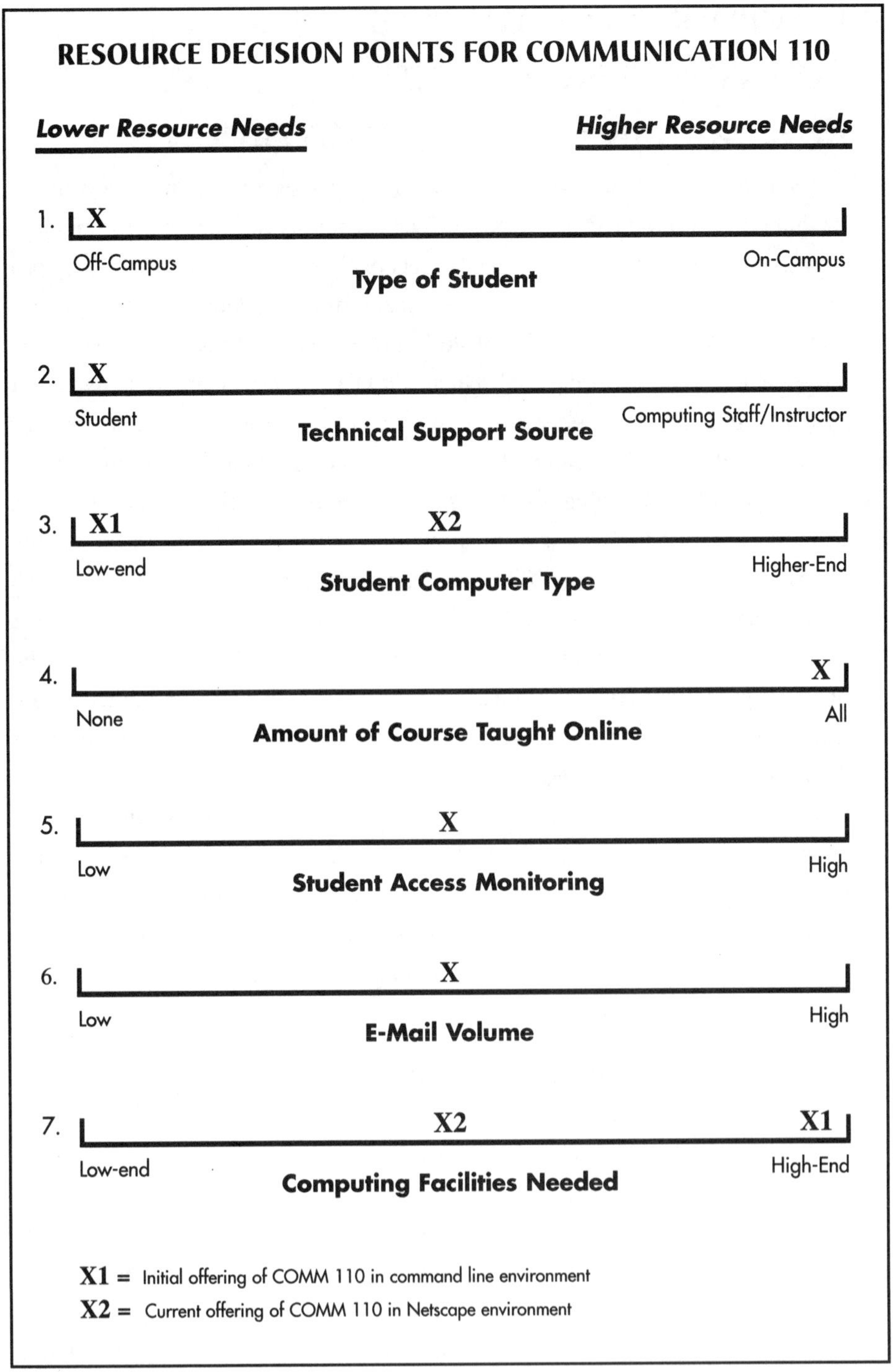

RESOURCE DECISION POINTS FOR COMMUNICATION 110

Lower Resource Needs   *Higher Resource Needs*

1. |**X**————————————————————————————————————|
   Off-Campus   On-Campus
                    Type of Student

2. |**X**————————————————————————————————————|
   Student   Computing Staff/Instructor
               Technical Support Source

3. |**X1**————————————**X2**————————————————|
   Low-end   Higher-End
              Student Computer Type

4. |————————————————————————————————————**X**|
   None   All
            Amount of Course Taught Online

5. |———————————————————**X**———————————————|
   Low   High
               Student Access Monitoring

6. |———————————————————**X**———————————————|
   Low   High
                    E-Mail Volume

7. |——————————————**X2**———————————————**X1**|
   Low-end   High-End
              Computing Facilities Needed

X1 = Initial offering of COMM 110 in command line environment
X2 = Current offering of COMM 110 in Netscape environment

Figure 5.1

Composing for the Web: Toward Interconnected Rhetorics in College Composition

Using the World Wide Web for content delivery—as a medium for posting and for linking to relevant material on the Web, if not as the medium for an entire course—is one of the most practiced facets of Web construction by college professors. Using the World Wide Web as a medium for student work—as a course bulletin board, journal, or archive of student pages and compositions—can appear more daunting, as requiring Web *composition* may require teaching Web *construction*. In addition to squeezing an introduction to hypertext markup language or the use of an HTML automation program into an already-crowded syllabus, instruction in Web building also necessitates articulating criteria for quality Web compositions, just as teaching standard written compositions requires articulating criteria for successful writing. The benefits of teaching Web construction early in the college curriculum, however, can be well worth the additional work in terms of connecting students with an audience outside of the classroom; encouraging revision; promoting critical, scholarly evaluation of one's own and others' work; and equipping students with a skill that can grow with them throughout their academic and professional careers.[2]

As I will argue in the first part of this section, quality academic Web construction is integrally linked to quality academic writing. Proficiency in both through use of the projector, as I will suggest in the latter part of this section, can thus result in better communication within electronic and print media, and in better coaching—on the part of students and instructors alike—on how to achieve effective communication.

Reading, Writing, and HTML?

Critics of incorporating Web construction in first-year composition courses often question the necessity of working HTML formatting and design into a course that already incorporates anything from library research and writing in a range of rhetorical situations to peer review of drafts, textual interpretation, and socialization into academic communities. The proposal for a computer-assisted version of Middle Tennessee State University's first-year writing course, for example (Clayton 1998), suggests that although word-processing, e-mail, and file-sharing in the computer-assisted classroom can nicely enhance the portfolio nature of the course, and although a course Web page can conveniently deliver content to students, the course will not be "a course that teaches HTML or encourages hypertext composition. Nor is the objective of this course to produce 'electronic portfolios' where the technology available overtakes and dominates the composition process." However, to allow, encourage, or require students to use the Web for research or for course materials, but to neglect to teach them how to contribute to that medium as part of their college education divorces students from the interactive potential of the World Wide Web and reinforces hierarchies in which the students are merely consumers of, rather than contributors to, electronic information. Moreover, in my own experience, sufficiently incorporating Web construction in composition courses has not necessarily meant allowing it to overtake the composition process, but rather, has allowed Web composition to complement and intensify more standard writing instruction.

I generally budget three hours of classroom time to introduce first-year composition students to basic Web construction: That is, HTML code (even though most students later gravitate toward automation programs such as Netscape Composer, Adobe Pagemill, and Microsoft FrontPage), basic manipulation of images and color, and techniques for uploading Web documents. This amount of time in a semester-long course has remained constant, whether or not the class has consistently met in a computer-assisted classroom, as did my 1995-1996 classes on argumentative writing at the University of Texas at Austin, or met in computer-assisted classrooms only for select class periods, as have my more recent classes on exposition and argument at Valparaiso University.[3]

In addition, I have built in readings on Internet use—for example, Deborah Tannen's "Gender Gap in Cyberspace"; Mark Surnam's "Wired Words: Utopia, Revolution and the History of Electronic Highways"; and Sven Birkerts, John Perry Barlow, Kevin Kelly, and Mark Slouka's *Harper's* forum, "What Are We Doing Online?"—into the courses' focuses on cultural studies and various types of literacies. Formal and informal assignments, as well, have encouraged research and evaluation of Internet media, for example evaluating real-time, synchronous class discussion (via the program InterChange in the Daedalus Intergrated Writing Environment) versus face-to-face conversation, performing a "scavenger hunt" of academic and recreational WWW sites, and evaluating select WWW sites or other virtual places. These are in addition to creating assigned WWW pages that students later present to the class with an LCD projector.

Like much other academic instruction, however, initial instruction in Web building—from the nuts and bolts of HTML code to the rhetorical finesse necessary for a successful site—requires students to spend additional time outside of class practicing, and ultimately mastering those concepts. Instructors in institutions with limited support for computer-assisted teaching may also need to budget extra office or lab hours for Web construction alone. Certainly, not every course has the time to build in Web construction or is an appropriate forum to do so. Even at institutions with limited resources, though, composition courses—particularly introductory ones—certainly can become such a forum. Increasingly frequent articles on Internet communication in composition readers, the inclusion of World Wide Web citation and style guidelines in writing handbooks, and the recent advent of rhetoric textbooks that include WWW composition offer evidence of the widespread place of WWW communication in writing courses.[4] But why squeeze a focus on new information technologies into already sound writing curricula and even risk, as the Tennessee proposal suggests, WWW construction overtaking the already difficult goals of first-year writing courses? Why not, at best, allow some other department to teach those skills in some other course?

One argument for teaching World Wide Web and more standard composition together is that successful academic writing requires several of the same skills as successful Web building: Integrating information from a variety of sources and expressing a rationale for the connections between them. Nowhere is the similarity between the two discourses more apparent than in the common pitfalls of beginning WWW construction, which quite uncannily mirror the pitfalls of inexperienced academic writing. An obliquely named or haphazardly placed hypertextual link that offers an audience little sense of where it is going quite accurately mirrors the improperly cited, unattributed, or poorly integrated quotation so common in first-year papers. Similarly, the irrelevant, improperly sized, or unauthorized image in a Web composition reflects a lack of expertise—on how to find and select appropriate images, manipulate them via HTML codes or size tags, and design and integrate images according to prevailing Web aesthetics— just as plagiarized or poorly synthesized material in research papers often reflects inexperience with integrating one's own words with outside sources.[5]

Rather than distracting students from sufficient practice in academic writing, discussion of effectively incorporating images or links to relevant material often reinforces discussion of effectively incorporating attribution phrases, clearly titling one's work, and respecting others' intellectual or otherwise copyrighted property. At one level, the technical discussion that teaching and learning Internet communication requires—the precision necessary for discussing HTML tags, passwords, file formats, and browser capabilities—can also augment more traditional instruction in drafting and revising effective prose. Moreover, whereas many students come to college pre-conditioned to believe that they are bad writers, and thus initially may be resistant to organizing paragraphs via topic sentences, beginning and ending sentences with dominant words and phrases, or drafting sufficient paraphrases, those same students are less likely to believe that they are bad Web designers, and thus may be more open toward learning new techniques and to giving and accepting advice in electronic media—skills that they can later transfer toward more traditional written communication.[6]

At another level, the discussion of audience inherent in teaching even the most technical of Web formatting dovetails nicely with the rhetorical dimensions of academic prose. Why one might choose relatively small, compressed images; break up long passages of text into smaller, interlinked pages; or link to, rather than embed sound or video files that automatically play ultimately depends on one's audience. This is the same reason that one might make a bridge to one's reader in an introductory paragraph, articulate one's thesis at an appropriate point in an essay, or fully develop one's paragraphs according to the expectations of an academic reader.

Success in Written and Web Compositions

The issue of audience leads to an additional set of reasons for integrating World Wide Web and more traditional compositions: Web communication's potential for infinite revision—a goal supported by recent decades' process-oriented models of writing instruction—and the intensity with which a WWW audience can elicit such revision. A student's brother pointing out a spelling error in a homepage, a friend from home asking for clarification about a point in an essay, or a proud parent referring co-workers to her daughter's successful paper on the Web can motivate far more effectively than the same comments from one's instructor. Often, as evidenced by my students' continued revisions to their sites, a Web audience continues to motivate after the semester ends. Having to choose one's most successful and appropriate compositions for a WWW readership additionally encourages self-assessment and consideration of one's audience, as well as how to present one's ethos online. Even students who come to college having already created Web pages in high school, on GeoCities or in some other setting, are often eager to create new ones or to refine and update already constructed material, composing for new audiences in the process.

Some success stories concerning Web construction, as evidenced by my students' evaluations of their work at the end of the semester, include the following testimonial:

> I feel proud of the work I did this semester. Especially not considering myself to have the best writing skills...my papers have shown a vast improvement. I have to say, however, that my Web page is what I am most proud of. When I first started to put it together, I had some difficulties, so I worked really hard on it and I think it looks very good. Also, I am proud that I was able to help others on this project. In this class, I have usually been the one to ask advice from others for my papers. But with the Web site, it was everyone coming to me to get help. It gave me a great sense of accomplishment that I was able to help others.

Still another student writes that in addition to coming to new insight and proficiency in use of the library as she drafted her research paper:

> Working on my Web site also helped me to get my hands dirty and not be so scared of the computer. Web building would not have been something I would have attempted or picked up on my own. It was helpful to learn html codes, the basis for the Web sites I visit weekly. I was proud to tell my Web builder little brothers that I, timid with computers, could make a smaller, simpler version of their creations.

Both students—the second, a confident writer at the beginning the semester but inexperienced with computers, and the first, an initially timid writer but less shy of Web construction—mastered incorporating attribution phrases and successfully synthesizing outside sources into their prose through a process similar to their creation of successful Web pages. Still other students from my 1997-1998 writing courses began to organize their sites (which began with a personal homepage early in the semester, to which they later appended two to three of their most successful or Web-appropriate papers) around a specific theme that resembles what David Joliffe identifies as an "inquiry contract": A series of compositions on a similar theme that emphasize revision and refinement of ideas not only within individual papers, but also from paper to paper.[7]

One student's initial homepage, along with her online papers on literacy in English and Spanish and on the conflict over a public beach, focuses on being

Puerto Rican-American. Another student's site and selected papers focus on an auto racing theme, and still another student's site showcases a motorcycle race in Ohio that he organizes with his father. That site includes a paper he is revising for publication in a motorcycle magazine and for an online promotion of the race. One student, who organized his site around papers on sports themes, maintains a frontpage on sport news that he updated throughout the semester, including daily predictions on the NBA draft.

Other success stories inherent in my students' Web compositions include connections to academic or extracurricular interests, "makeovers" of pre-existing Web compositions, and projects in community service or other fields that students have undertaken through Web construction. An electrical engineering student, for example, included on his site an image he had drawn with AutoCAD. Another student linked to the download site for a program she had been using in one of her civil engineering classes, while still another student linked to his football statistics, which the athletic department had posted online.

All of the above sites also experimented with various types of Web design, with speeding up the time it takes to download long pages in mind. Additionally, two students who had come to class with sites that they still maintain on GeoCities gave their sites extensive makeovers, with one student updating her site for a collegiate audience and streamlining its organization through a table-based menu, and another student excising a distracting background, embedded sound files, and a minimally related list of "hot links" in favor of a Javascripted, more tightly organized compendium of interests that includes an index of all of his work from the semester. Rather than considering the Web projects as a reiteration of what they already knew, the students—just as experienced writers often accomplish in writing courses—took advantage of the assignment to expand their expertise. For both students, such exercise has paid off in the form of work and community service opportunities. The first student is teaching Web construction to elementary students through area boys' and girls' clubs, and the second student is creating Web pages for local businesses. Similarly, other students have moved on to creating Web pages for their sororities or fraternities, area music groups, literary societies, or churches.

Forging such cognitive and experiential, as well as hypertextual, links to one's areas of expertise illustrates academic maturity and, I hope I have argued, is dependent on not only successful Web, but also written composition. Meeting the challenge of new cultures that students bring to the classroom, however, cannot end in first-year composition. Building a critical mass of students who are literate in composition for the World Wide Web builds an expectation that upper-division courses will build on that knowledge. Ann Reiser's education courses, discussed in the following section, offer one example of integrating more advanced courses across the curriculum. Course sites such as Rice University's "Galileo Project" or Clarke College's "The Last Days of Socrates" additionally offer salient examples of course projects that draw expertise from across the university (for example, site design and advice on art history from arts departments, assistance on programming from computer science students, and research from appropriate philosophy or history or science students). Course sites—for example, a literature site that puts out-of-print (and not copyrighted) poetry online—can allow students to research and publish the literary and historical context of the work, while perl scripted Web forums on the site can allow students or other interested visitors to annotate, question, or discuss unfamiliar words and phrases.

Such projects, if overly ambitious for period courses that are offered only infrequently at small institutions, can profit greatly if shared over cyberspace with similar classes at other institutions. Whether used for individual classes, multiple sections of a course, or in conjunction with courses at other universities, the World Wide Web can forge and enhance, rather than replace, the essential relationship between writer, audience, and text. In so doing, Web use can thus accommodate the changing contours of communication while honing and developing them throughout the college experience.

PERSONALIZED ELECTRONIC TEXTBOOKS: SHAPING A NEW LEARNING ENVIRONMENT FOR PROFESSIONAL EDUCATORS

Recently it became startlingly clear to me that the textbook as the university education professor knows it might soon cease to exist. Recent advances in telecommunications technology have made it possible to deliver over the Web an interactive learning environment.

Future teachers need to understand the deep impact technology is having on society as a whole, how technology has changed the nature of work, communications, and our understanding of the development of knowledge. Future teachers must recognize that information is available from sources that go well beyond textbooks and teachers—mass media, communities, and so forth—and help students understand and make use of the many ways in which they can gain access to information. Future teachers must employ a wide range of technological tools and software as part of their own instructional repertoire.

Although many teacher-education programs offer a course in instructional technology, some are unable to do so or only offer a one-credit course that is expected to cover all of the technology resources needed to equip future teachers to enter the 21st century. However, national organizations such as the International Reading Association, the National Council of Teachers of English, and the National Council for the Accreditation of Teacher Education strongly urge or require that certain competencies be included in all teacher-education major areas.

As a university professor involved in instruction of future teachers, I am observing a new generation of students entering the university. They are arriving already imbued with some background and motivation and are ripe for guided inquiry, ready for interpretation and collaborative construction of knowledge. Students are coming to the university having experienced television, videotape players, computers, Web TV, and video games.

It is within this context that I began the integration of Web-based instruction into my university classes. I wanted future teachers to have a model for how Web applications can enhance teaching and learning. I wanted them to be able to use Web development tools as HTML editors, and to use graphics and communication software. I wanted the future teachers to use the Web to complement their classroom instruction. I wanted all of this to happen with a three-credit language arts course that is a part of a professional block program.

Web-based instruction has allowed me to create an interactive learning environment, which became a personalized electronic textbook made up of links to sites on the World Wide Web. Opinions differ about the future directions that education should take regarding the use of textbooks and the value of textbooks to communicate knowledge to students. The Web-based instruction allowed me to tailor information to a specific topic or area of research.

Drucker (1995) states, "It is a safe prediction that in the next 50 years, schools and universities will change more and more drastically than they have since they assumed their present form more that three hundred years ago, when they reorganized themselves around the printed book" (p. 79). I believe these changes are being driven by a number of forces, such as the following:

1. Microcomputers

2. CD-ROMs

3. Interactive distance learning and virtual reality

4. The demands of federal/state politicians and business leaders

5. The demands of our knowledge-based society

It appears unlikely that the pencil-and-textbook traditions of yesterday's classroom will continue to hold such a dominant position in American education. Ornstein (1995) speculates that "the textbook as we know it is doomed to obsolescence" (p. 266). The standardized textbook pales in comparison to the recommended minimum standards for classrooms of the future, which include a television and VCR with cable and down-link access, World Wide Web access, a multimedia computer, and an electronic presentation system.

Over the past year, I have developed an online language arts class entitled "Learning Together through Inquiry." Telecommunications technology has made it possible to create a personalized electronic textbook made up of links to sites on the World Wide Web. Pedagogically, students become active learners and work with information sources far more current than those typically provided by the textbook. At the beginning of the course, students must quickly and effectively learn how to use the World Wide Web and research cutting-edge information systems topics. The students are given a period of time for basic Web construction and observe the personalized electronic textbook site being constructed during each class session. Each week the students are challenged to use this site to gain additional knowledge regarding a specific topic.

I have found the World Wide Web connection to be a positive part of the language arts class. The students would comment in the course evaluations that they became more familiar with the World Wide Web as a tool for research, education, professional development, entertainment, and personal growth. Many of the students felt comfortable in developing and implementing classroom Web sites in their future classrooms. Since Web construction is such a successful part of the course, I will continue to use the online language arts class, "Learning Together through Inquiry," while also challenging the students to develop their own Web-based professional portfolios.

The technological advances and demands for greater accountability and increased interest in performance assessment will push the students to another level of professional development. The portfolio will provide tangible evidence of the wide range of knowledge, dispositions, and skills that they possess as aspiring professionals. I hope to base the organization and content of their professional portfolios on the Interstate New Teacher Assessment and Support Consortium (INTASC) standards. These ten standards are written in terms of performance and knowledge and serve as a core of expectations for all teaching. A list of items to include in a Web-based portfolio will be provided at the beginning of the course. These items will relate specifically to the online language arts class. I am looking forward to this new teaching challenge as I push forward in rethinking traditions and meeting institutional challenges.

References

Barlow J., Birkerts S., Kelly, K., Slouka M. (August 1995). What Are We Doing On-line? *Harpers*, 35-46.

Brent D. (Spring 1997). Rhetorics of the Web: Implications for Teachers of Literacy. *Kairos: A Journal for Teachers of Writing in Webbed Environments, 2.* Available online at http://english. ttu.edu/kairos/2.1/features/brent/bridge.html, specifically "Co-opting the Clickable Classroom," available at http://english.ttu.edu/kairos/2.1/features/brent/ coopclic.htm.

Burow-Flak E. (forthcoming). From Castles in the Sky to Portfolios in Cyberspace: Building Community Ethos in First-Year Rhetoric and Composition. In Gruber S.(ed.) *Weaving a Virtual Web: Practical Approaches for Teaching with New Information Technologies.*

Clayton M. (1998). Computer-assisted, Portfolio-based Composition: The Next Step for Freshman Composition at MTSU. *Proceedings for the 1998 Mid-South Instructional Technology Conference.* Available at http://www.mtsu.edu/~itconf/proceed98/mclayton.html.

Drucker P. (1995). *Managing in a Time of Great Change.* New York: Truman Talley Books.

Howard R. (1995). Plagiarisms, Authorships, and the Academic Death Penalty. *College English*, 57, 788-806.

Hull G., Rose M. (1989). Rethinking Remediation: Toward a Social-Cognitive Understanding of Problematic Reading and Writing. *Written Communication*, 6, 139-154.

Joliffe D. (forthcoming). *Inquiry and Genre.* Boston: Allyn and Bacon.

Ornstein A. (1995). Curriculum Trends Revisited. In Ornstein A., Behar L. (eds.) *Contemporary Issues in Curriculum*, Boston: Allyn and Bacon.

Panici D. (1998). New Media and the Introductory Mass Communication Course. *Journalism Educator*, 53, 52-63.

Soules A., Adams E. (May/ June 1998). Classroom Technology: A View from the Trenches. *Educom Review*, 50-53.

Surnam M. (1995). Wired Words: Utopia, Revolution and the History of Electronic Highways. Available at http://www.web.apc.org/~msurman/wiredwords.html.

Tannen D. (1996). Gender Gap in Cyberspace. *Writing Lives: Exploring Literacy and Community.* S. Garnes, D. Humphries, V. Mortimer, J. Phegley, and K. Wallace, eds. New York: St. Martin's Press, 1996.

Williams J. (1995). *Style: Toward Clarity and Grace.* Chicago: University of Chicago Press.

Endnotes

1. In its current manifestation, COMM 110 requires that student use Netscape 4.05, either on their home machine or in campus labs. Students are responsible for establishing working e-mail accounts, either with the university or another ISP.

2. For suggestions on promoting understanding of a WWW audience as part of instruction in WWW construction, see Burow-Flak, "From Castles in the Sky to Portfolios in Cyberspace: Building Community Ethos in First-Year Rhetoric and Composition," forthcoming in *Weaving a Virtual Web: Practical Approaches for Teaching with New Information Technologies.*

3. For the WWW pages of my fall 1995 and spring 1996 rhetoric and composition courses at the University of Texas at Austin, consult the following WWW addresses: http://www. cwrl.utexas.edu/~betsyb/classfall95/ and http://www.cwrl.utexas.edu/betsyb/classspring96/. For the WWW pages of my fall 1997 and spring 1998 exposition and argument courses at Valparaiso University, consult the following WWW addresses: http://www.valpo.edu/home/ faculty/bflak/E100fall97/ and http://www.valpo.edu/home/faculty/bflak/e100spring98/.

4. For composition textbooks that discuss WWW building, see Anderson, Benjamin, and Paredes-Holt, *Connections: A Guide to On-line Writing* (Allyn and Bacon 1998); Clark, *Working the Web: A Student's Guide* (Harcourt Brace 1997); Condon and Butler, *Writing the Information Superhighway* (Allyn and Bacon 1997); and Vitanza, *Writing for the World Wide Web* (Allyn and Bacon 1998).

5. For a discussion of "patchwriting" in Web compositions as it mimics a similar format in basic writing, see Brent's section on patchwriting in "Rhetorics of the Web" at http://english. ttu.edu/kairos/2.1/features/brent/patchwri.htm. For Rebecca Moore Howard's definition of "patchwriting," see "Plagiarisms, Authorships, and the Academic Death Penalty." For a discussion of such a technique as an intermediate stage of the writing process, see Hull and Rose, "Rethinking Remediation: Toward a Social-Cognitive Understanding of Problematic Reading and Writing."

6. For salient, but very technical suggestions on beginning and ending sentences with dominant, rather than weak words and phrases, see Joel Williams's *Style: Toward Clarity and Grace* (University of Chicago Press 1995). For discussion of fears of committing plagiarism that can stunt risk-taking and legitimate writing processes, see Howard in list of references.

7. For a textbook introduction to the inquiry contract, see Joliffe, *Inquiry and Genre* (forthcoming from Allyn and Bacon).

CHAPTER 6

Distance Education 2010:
A Virtual Space Odyssey

Steffen W. Schmidt
University Professor of Political Science and International Studies

Iowa State University
Ames, IA

INTRODUCTION

In the summer of 1998, a petition was drawn up by 850 University of Washington (UW) professors and sent to the governor. The professors were worried about the "... enthusiasm that Governor Gary Locke and his advisers are showing for instruction via CD-ROMs and the Internet." This got national attention through a story in *The New York Times* special section "Circuits" (June 18, 1998). The professors decried the diversion of money from "live" education to "techno-substitutes." This was only one of many anxious responses from academics throughout the United States over the past few years on the issue of how to promote technology and education in higher education. The virtual classroom and teaching at a distance with technology is not a *future* threat, as the Washington professors seem to suggest. The virtual classroom is here[1] and it does not have to be a threat. It can be a great opportunity.

The UW professors clearly assume that education is what I have called "geocentric." It occurs and is available within a geographic space-the local and physical classroom (the University of Washington, the city of Seattle or Tacoma, Washington State University, the United States). It also assumes that public policy makers such as Governor Locke[2] and his associates can, through legislation, prevent the penetration of distance learning to the lovely UW campuses and prohibit current or future UW students from enrolling in courses taught at a distance. Neither is correct.

In fact, a very important recent article in the *AAHE Bulletin* of the American Association of Higher Education (AAHE) concludes that distance education and the virtual classroom is already a thundering reality:

> Quite suddenly, in just two or three years, American higher education has come face-to-face with an explosive array of new competitors. On campus, the surest conversation-stopper today is University of Phoenix. To some academics, Phoenix looks like the first-sighted tip of an iceberg. But it probably won't be the one that sinks whole ships. Bigger bergs are forming.[3]

In Wisconsin, there are already over 100 out-of-state degree providers. The same is true of most states. One would not be shocked to discover that UW students are already taking courses at a distance or from a host of diverse, interesting, flexible, agile providers both before coming to college, while in residency at UW, and certainly after they complete their degree as part of lifelong learning needs.

THE DISTANCE LEARNING REVOLUTION

Distance learning is nothing short of an unfolding (albeit gradual) revolution that will deeply transform post-secondary education. Britain's 168,000-student Open University is already a global player in this area. The University of Phoenix is the largest private university in the United States, with over 40,000 students and no campus. Dallas Community College telecourses have been a big player for many years and my American government textbook has been the official book for their American government course. I've lectured to and visited informally with hundreds of alumni and instructors of this course all over the United States. My general impression is that these telecourses made it possible for bright but busy people to continue their education. Most have eventually taken the time to complete their degree in a "traditional" college setting. Harvard University, Duke, the Western Governor's University, the California Virtual University, and the Colorado Community College system (which now offers a two-year degree entirely in virtual format) are also notable. In Canada, Simon Frasier, University of British Columbia (only a stone's throw from Seattle and the concerned UW professors), and other institutions are aggressive distance-education pioneers.

The Reasons for the Revolution

In a recent report, consulting firm Coopers and Lybrand predicts a huge shift to virtual education for the basic college courses. The report states that "Software will serve an estimated 50% of the total student enrollment in community colleges, as well as an estimated 35% of the total student enrollment in four-year institutions."[4] This revolution has several roots.

First, the cost of a college degree is escalating by a factor of at least two greater than inflation. Escalating expenses have created enormous pressure for cost containment.

Cost is driven by several factors:

- The enormous expense of brick and mortar investment in traditional campuses and classrooms (and the maintenance overhead for the infrastructure)

- The expense of maintaining a huge, permanent workforce of professors, often for the life of the employee (tenure)

- The desire to provide substantial amenities, including housing, recreation, and a huge service infrastructure-counselors, health services, security, parking-that parallels public services available in the community

- The pressure and costs for winning varsity sports teams

Second, a predicted explosion of college students at the beginning of the new millennium requires that new ideas and innovative strategies be implemented to accommodate this surge in demand.

Third, the work and life pressures on today's student make a full-time residential experience extremely difficult for many. Convenience and the need for time flexibility are a premium factor in college or class decisions. Programs that allow students to complete or continue the educational experience while working, caring for children or other dependents, and living in a convenient location (not on the campus or even in the city where the university is located) are very powerful attractions to distance learning.

Fourth, knowledge is changing so fast that new academic programs, courses, and content will need to be created, delivered, enriched, and modified at unprecedented speed.

Fifth, a substantial component of traditional university education does not meet the qualitative needs or expectations of students. The research university, in particular, has shifted the promotion, tenure, reward, and incentive system for colleges, departments, and individual faculty from teaching to research. Thus, for many students, the attention they get and the quality of general education classes in many cases leaves a great deal to be desired. It is hard to argue that those classes of 500 or 1,000 are "traditional," face-to-face education. It is also a sad reality (as I've personally witnessed in my travels around the United States doing research on distance learning) that often professors have very little time or interest to give their students in office hours and informal interaction.

At most research universities graduate teaching assistants (TAs), not professors, have responsibility for a substantial and growing number of classes. In many schools, TAs teach over 50% of classes in some programs. Many assessments, including a close reading of the *Chronicle of Higher Education*, indicate that students are dissatisfied with the level of attention, mentoring, career guidance, and one-on-one interaction they receive from some of their professors. Distance education and virtual classrooms may be a perfectly acceptable alternative. After all, distance education encourages the student to contact the professor frequently through e-mail, telephone, real-time discussion, chat rooms, and other means.

Sixth, every discipline is now deeply affected by information technology. Every student should be exposed to computers and technology and should have at least a portion of their education take place through this medium. In private business, in K-12, and in medicine, distance technology is becoming a very common medium for delivery and interaction. One could argue that it is a good thing for students to understand, and become adept and comfortable with this medium.

Seventh, technology has an attraction in its own right. The World Wide Web in particular is affording users a level of graphic presentation, speed, flexibility, and excitement that most classroom instructors are unwilling or unable to match.

THE CHANGES IN ACADEMIA

Some critics of higher education argue that, unlike the changes that are sweeping the business community in the United States, higher education has essentially remained in a steady state for fifty years. They aver that, like it or not, the changes taking place in the private sector (consolidation, buy-outs, speed, quality/price formulas, "lighter and faster," trendy, reengineering, "right-sizing," just-in-time, total quality improvement, and so forth) are a reaction to market forces, cost, technology, and rapidly changing demands. Post-secondary education is not very agile, they say. Many argue that post-secondary education is also riddled with redundancy, unnecessary duplication, waste, and inefficiency, especially at state universities, which receive substantial taxpayer subsidies.

I think this critique is overstated. In fact, change has been constant in post-secondary education. All of the following are major changes in the old Ivy League model:

- The advent of community colleges (which was and still is much feared and/or ridiculed by many in higher education)

- The delivery of outreach and extension courses and programs (Extended and Continuing Education)

- Night classes and weekend MBAs

- The creation of McCourses—"fast courses," often compressing a whole semester or year into a few weeks, usually in the summer

- The rise of what I call the "MegaplexClass"—courses with huge numbers of students in large auditoria

- The practice of hiring temporary faculty who are paid for "piece work" (one course at a time)

- The outsourcing of the first two years of college to community colleges

All of these forces are now converging and will require a new philosophy and a new management of post-secondary education. The important questions are: How will this revolution unfold? Who will steer it? Who will control it? How can it be directed constructively so that American post-secondary education will remain a world leader?

For most colleges and universities the strategic policy challenge is to determine their strengths and weaknesses, and to project the opportunities and threats that lie ahead (the so-called SWOT Matrix analysis). The task is to build on the strengths; mitigate the weaknesses; and assess how technology, distance education, and the virtual classroom can best be embedded in their strategic plan. Reducing the negatives of competitor challenges (threats) and enhancing the opportunities including technology and the virtual classroom is extremely important (see Table 6.1 on page 80).

In the feedback we've received from interviewing administrators, faculty, and students around the county, there are many quantifiable factors that must be brought to bear on addressing the SWOT question.[5] Virtual classrooms and distance learning are not an isolated variable but one of a host of considerations that determine student selection of a college or a class. These include quality of dormitories, campus safety, access to computers and a wired campus, the quality of teaching and teaching evaluation, the library, the climate and geographic location of the campus, diversity of students, the quality of the community (is it student friendly, interesting, lively, fun, challenging).

Table 6.1 SWOT Matrix for Distance Education (Strengths, Weaknesses, Opportunities, Threats)							
Present	Quality of Faculty	Reputation or Ranking of Program	Cost	Funded Endowment	Enrollments	Facilities on Campus	Location & Climate
Strengths							
Weaknesses							
Future							
Opportunities							
Threats							

Note: This is a traditional SWOT matrix applied to the class, program or institutional capabilities now and the future environment relative to competitors. These seven constitute the basic assessment categories. You will need to insert a whole series of additional variables relevant to your institution or program.

POTENTIAL PROBLEMS OF DISTANCE LEARNING

Another question we should be asking is: How can distance education and the virtual classroom improve rather than damage higher education? The fear of "Wal-Martization" or "McEducation" is understandable. I believe in quality as much as the next professor does. But virtual education, if it becomes a problem, is a potential *future* problem. It is solving the real and *current* quality problems in higher education (not the anticipated potential future problems) that should be the highest priority of the American Association of University Professors (AAUP) and other quality watchdog groups. Moreover, the demand for fast, convenient, reliable, and inexpensive products is as much a reality in the world of academia as it is in retail or food. Like it or not, most people cannot afford to shop at Abercrombie & Fitch or eat at the Four Seasons restaurant.

Another fear is that education with technology diminishes the quality of learning. I can't objectively address the valuable and wonderful intangibles of leaving home, moving to a campus, living with other students, interacting with new people, and, in general, having the college experience. I can, however, cite very impressive literature that suggests that distance learning is at least as good, and sometimes better, in terms of learning and retention. Keene and Cary indicate that "when compared with students taught conventionally, students who received the distance learning instruction evinced superior knowledge of the subject matter at the end of instruction."[6]

A similar testimonial comes from the field of engineering where Wergin and Haas[7] found that "There are generally no significant differences between remote and on-campus students in academic performance, and in some cases the TV students outperformed those on campus." Delonghy[8] reported that students found that courses taught through computer-based conferencing: "(1) improved the educational quality of their courses; (2) made access to education more convenient; (3) involved them more actively in the classes; and (4) improved the access to their professors." Thomas Cyr, one of the gurus of technology and distance teaching, cites a long trail of similar studies in his excellent book *Teaching at a Distance with the Merging Technologies*,[9] most of which find quite positive results and high student satisfaction with virtual classes.

THE POSSIBLE BENEFITS OF DISTANCE LEARNING

By shifting functions and economic incentives and adding strategic distance education components, universities can in fact better concentrate on their primary mission. For Carnegie I institutions, that may be research and advanced learning. For small liberal arts schools, it may be teaching, close interaction with students, and mentoring. For community colleges, it is something else-highly focused specialty programs and training on the one hand, entry-level courses or even remedial courses, on the other hand.

A more rational division of labor-in other words, specialization-could help relieve the research university of the burdensome, inappropriate, expensive, and corrosive "service" functions (such as remediation, and broad survey courses) that now occupy a preponderance of departmental time, resources, classroom space, and professors. Many of these courses can be delivered more effectively, more cheaply, and with similar learning results at other institutions and/or by using technology.

In reality, the issue is not as crisp as the virtual versus traditional classroom dichotomy suggests. The role of technology in the classroom is significant, regardless of whether it is a real-time, synchronous, on-campus class or a virtual classroom. There are many devices used in the electronic classroom that deliver all of the following to a large screen in the classroom, normally through a networked video projection system: VCR tape clips; laptop presentations like Power Point; CD/ROM accessed material projected from the laptop; live World Wide Web sites driven from a high speed Ethernet line; laser disks; and desktop video projection (we call them Elmo's), which allows one to place paper, a book, or other objects under the video arm and then project them to the large screen. Slides, maps, reel-to-reel film, and overhead transparencies are also part of classroom technology. Come to think of it, so is the mouth and chalk!

Note that in this example the classroom is rich in technology. It's what I've called an "e-classroom".

However, it is *not necessarily* a virtual classroom. The professor and students are in the same room at the same time and are physically in the same place. Incidentally, I use all of the above in carefully scripted combinations to make the classroom alive, exciting, more powerful, and fast-paced. Table 6.2 (see next page) explains how I script a typical class.

Thus, distance education or the virtual classroom can simply mean taking a "real" or traditional classroom, such as the one above, and then making it available at locations other than the actual classroom at the same time (this would be synchronous distance learning). Alternatively, the material can be "captured" electronically and archived on a Web site. Students both on campus and at a distance can then access the material at any time after the class is over. (This is asynchronous distance learning).

With emerging technologies (and with the "merging" of these technologies), it will be possible for students at a distance to participate in either the real-time class or the "time warped" virtual class. Small, high-powered videocams will attach to the monitor of their computer or laptop and sensitive microphones will be used to send visual and oral comments from remote-site students to the class, or used to add their comments to the virtual Web site for review later.

Table 6.2

Story Board-Political Science-215-American Government-Fall 1997-Schmidt

Chapter #:   Date to be Delivered: Topic:

Minutes	Real-time	Material/Activity	Comments
5	.... 9:30	Music-in. I play a CD to help students transition into my class. Each time a different genre of music. With a graphic, looped power point slide show using photographs that set up the theme for the day.	All of the class activities can be videotaped, digitized and placed on WWW server for students enrolled at a distance.
5	9:30-9:35	Class management and housekeeping. Announcements, reminders etc. With overhead transparency and chalkboard.	
5	9:35-9:40	Power Point outline of unit. Key points in the textbook. Terms or data.	Microsoft PowerPoint
20	9:40-10:00	*Mini- lecture #1* Current case study that reveals the timelessness of readings and topic. Tech: Live WWW material on Ethernet connection or Web Whacked off-line.	
10	10:00-10:10	Interactive activity in which students make a decision linked to the case study.	(Paper interactive handouts). Points are awarded for students in class this day or submitting from a remote site via fax.
10	10:10-10:20	Discussion of activity and conclusions students draw about this case. Students come up to front of class and use laptop or graphic tablet for their presentation. (Image in Comments column at right shows an example of a Web presentation design session.)	
20	10:20-10:40	*Mini-lecture #2* A second important cluster of material analyzed or explained. Tech: CD/ROM or Laser Disk projection on large screen with my oral lecture.	
3	10:40-	Music-out. To help mental-exit transition from class. Closure. Power point slide show of "coming attraction" (next class).	Sen. Kassebaum guest lecture.

Distance testing is already being done successfully through partnerships (and fee for service) in which colleges or other education providers physically close to the distance student serve as monitored testing sites (community colleges, Sylvan Learning Centers, U.S. military training, and education facilities). Students then can take essay or multiple choice exams or be tested orally online with appropriate supervision. I already use computer labs that I monitor to let students write their essay tests on campus. I've also successfully used online testing.

In my classes, I make material available for distance learning use even though all of my students in any given semester may be enrolled on campus and physically come to class. I see them as "distance learners" because they can access material from their dorms, apartments, the library, a computer lab, or, if they need to leave campus, from any PC anywhere in the world. I had a student last semester who suffered an illness and had to go home. He was able to continue two weeks of class at a distance from his computer in his bedroom and then return to campus, resume synchronous classes, and barely miss a beat.

If students in China or Germany wish to take a class of mine, they eventually will be able to do so, too. Their distance from me in my real class will be imperceptible because they will appear on screen, communicate, take a quiz, participate with a learning team in my class, and virtually be no further away than the 400th student in the top row of the auditorium. The concept of "distance" will then change dramatically! Most likely my distance education students will be largely a class that also has a substantial, majority on-campus enrollment and meets in a "real" classroom. This would, of course, not be possible for many classes where a laboratory or specialized equipment is needed. It might also be difficult for certain courses where group activities are necessary, although I've found in the past that if the technology is properly structured, groups of students at physically separate locations can work as teams.

The fact is that a professor can move a class from no technology, to low technology, to an e-classroom, to a virtual classroom (and do this synchronously to asynchronously), as they wish. Table 6.3 on page 86 illustrates the dynamics of these options.

THE APPROACH TO DISTANCE LEARNING

Ideally, of course, distance classes are carefully designed for this special purpose. Much as a symphony orchestra consists of many instruments-a conductor, a score, publicists, promotion and marketing experts, support staff, and technical persons-so a virtual class is also best when it is richly produced. My own preference is for universities to have a very structured approach to distance education with a clear overall plan. I also advocate the creation of a Virtual Course Design Studio, which will provide the entire tech support, bells, whistles, and production excellence so that the expert professor can deliver a truly high-quality course.

However, and to follow the musical metaphor:

> Your use of technology should be guided by a very simple paradigm. If it helps you teach more effectively, use it. If it doesn't, don't, but don't fault others for using it because it might work for them. Technology can do some wonderful things and give your students learning opportunities that are totally impossible without technology. Think of technology forms (as well as other instructional methods) as learning instruments that you orchestrate to produce the symphony that is your course. Use the flutes or trombones only when they are appropriate, but don't write them out of your score simply because you are skeptical of their value![10]

I say Amen to that.

As a colleague pointed out to me as I was completing this chapter, some academic fields such as student counseling (his area) can only make limited use of the virtual classroom because of their inherent nature. He indicated that Harvard University gives a one-year degree in this area, which several colleagues of his obtained and which they considered of very poor quality because the hands-on, human interaction factor was missing. I would go further and say that most of us are still skeptical of full-blown virtual degrees. However, we are optimistic about distance education that is used intelligently.

Critics of distance education and the virtual classroom are often, from our research, those who are least familiar or comfortable with instructional technology or those in programs or at institutions where technology is outdated or not available. Many are understandably frightened by the potential threat to quality academic jobs and careers. (In reality, thousands of new jobs in academia are being

Table 6.3
GeoCentric and DigiCentric Instruction

GeoCentric (Traditional)	
Example: Conventional lecture-style class	Example: A class based on programmed instruction manuals (self-paced and self-corrected)
Synchronous Example: 2-way video conference-style class	**Asynchronous** Example: A computer (WWW)-managed class replete with listserves, electronic files, chat, audio/ video streaming, on-line testing, etc.

DigiCentric (Virtual)
Two major variables define the twenty first-century classroom. • Is it centered on a physical classroom where the instructor meets students face to face and where students come to campus for classes (Geocentric) or is it virtual (DigiCentric)? • Does it occur simultaneously for all students (Synchronous) or does each student access it at their convenience and perhaps even repeatedly if they so choose (Asynchronous)? • The simple model above presents four choices in the combination of the x and y axis variables although in reality much more complex permutations are possible.

created by the digital and virtual education revolution.) Others worry about the issue of intellectual property rights. Many worry about the integrity and fundamental nature of a university education. All of these concerns and more must be taken into account as we proceed down this road of virtual education.

Technology will soon make it possible to integrate synchronously or asynchronously students and instructors at disparate locations in such a way that interaction is quite excellent. After all, a student in the back of an auditorium in a class of 400 cannot be seen or heard very well and certainly must overcome a massive psychological barrier to speak up. A student in a distance education class of 400 others, sitting in front of a high quality monitor with appropriate video and sound, to both speak and be seen and heard by the professor and the other 399 students at distant sites, may be both more visible and audible and certainly would feel less threatened to speak out during discussion period.[11]

University of Washington Professor Alan Borning worries that students will be "… learning from CD-ROMs off in their garages" and, thus, not interacting with other faculty and students. This is an understandable concern but we hope it represents the worst-case scenario, one that none of us should or would tolerate, including students who are generally smart consumers. Rating systems and "brand names" will, in the near future, shake out the good and the bad programs.

The question, "Who 'owns' the course and the intellectual property?" is a serious issue. Will the university offer a class once the professor has been fired, retired, left for another institution, or died? My view is that this will and must be clarified through intellectual property contracts that will guarantee the rights of all parties. For example, if a university expends $100,000 for the production of video tapes of scientific material to be used in conjunction with a class, one could make the case that this material might remain in the chemistry or biology department of the university for use by all faculty in the future.

However, much as copyright laws protect a professor who writes a textbook, distance education material that has the feel and quality of original intellectual material should be protected in the same way. In my experience this will not be a huge issue since most lectures or discussions in classes hardly qualify. In-class activities, such as discussions or lectures, often are obsolete at the end of the semester. The same is true of video or Web-centered courses. The worry that 10 years later the university will exploit this material is either a red herring or a straw man. The stuff is not that good and becomes obsolete quickly.

Already movies, videos, software, photographs, music, graphics, and other material is copyright protected and cannot be used without permission in distance education courses. Professors who develop such material or original text for courses will need to copyright them. The university or other course developers will then need to get permissions, user rights, and probably pay a fee for future use. Professors will simply need to contact their local copyright attorney and become more savvy entrepreneurs of their own intellectual property.

Does virtual education raise serious issues about post-secondary education? Of course it does. Although I am an enthusiastic supporter, consultant, and practitioner of virtual education, we need to take into account best practices for distance learning. Some of these are noted in Table 6.4 on the next page.

Table 6.4 Good Practice and Major Instructional Delivery Systems at a Distance

	Print	Correspondence	Internet	WebSite	Computer Mediated Conference	Audio	Video & Laserdisks	Real-Time Video	Compressed Video
Faculty-student interaction									
Collaborative learning and teamwork									
Active learning									
Efficient use of student time									
Rich and rapid feedback on assignments									
Challenging assignments that communicate high expectations									
Respect for diverse talents and ways of learning									
Higher order and critical thinking									
Comfort and familiarity w/information technology									
Use of the scientific method									

This matrix allows you to systematically track the use of instructional delivery methods and their relationship to good practices in education. Chickering and Gamson, "Seven Principles of Good Practice in Undergraduate Education." , *Wingspread Journal*, 1987, & Chickering and Ehrmann, "Implementing the Seven Principles: Technology as Lever", *American Association of Higher Education Bulletin*, October, 1996. I have added several good practices to the original seven. *Internet Workshop Exercise*: Team activity. Each team identify a use of the Internet and World Wide Web for each good practice. Report back to the group. Discussion.

Some Guidelines

As we assess how to best juggle traditional education and the virtual environment, the following six factors are useful guidelines:

First, I believe that university professors are the keys to successful and high-quality distance education. They are the repositories of both the knowledge and the methodologies that lead to successful research and to well-trained and well-educated scientists, artists, leaders, thinkers, and citizens. Good distance education—no, great distance education—will require that the great professors and researchers are full stakeholders and participants in the process.

Second, academics need to accept the fact that distance and virtual education, as well as technology in the classroom, are not passing fads but a revolution. Whether they like it or not, it is here and the consequences of this transformation will forever change higher education.

Third, instructors should introduce technology into their teaching as appropriate. However, we should realize that students have always demanded relevance in education. The curriculum of 1998 is in good measure the result of pressure and demands coming from students. Thus, students have always been "consumers" or "customers," even though we've avoided referring to them as that in the past. Today's students are socialized in a world of images, sound, graphics, and very fast movement. University teaching cannot escape this reality. My own American government textbook was the first four-color, heavily illustrated book in this field. We have not compromised quality or substance but we deliver much of this information in new ways that connect better with students.[12]

Fourth, policy makers must understand the first point-professors and discipline experts are the crucial component to high-quality distance education. Without them, it is just a delivery vehicle. It is the content that matters. Thus, universities will need, more than ever, to compete for the best-trained and most entrepreneurial faculty. These professors, as always, will be vital to the overall quality of on-campus, off-campus, real and virtual, synchronous and asynchronous teaching.

Fifth, state legislatures and governors can and will play a powerful leadership role in directing resources, regulating them, offering incentives, and pushing for structural innovation to make distance learning a success. This is especially crucial for publicly funded universities. They can help forge real partnerships and valuable strategic alliances among the private sector, academia, and government in this area.

Sixth, distance education will be one of the determinants of how competitive American higher education remains and whether the United States will continue to be the world leader.

CONCLUSION

Clearly, a lot is at stake here. Above all we must avoid making the virtual education, electronic teaching revolution a zero-sum battleground between traditional academia, cost-conscious government, profit-seeking private education companies, eager students, and anxious parents. Everyone can be a winner if enlightened leadership prevails.

Endnotes

1. A very interesting article on this is "Drive-Thru U" by James Traub, in The New Yorker magazine, October 20 and 27, 1997, pp.114-123.

2. Governor Locke appointed an interesting "2020 Commission" made up largely of business people, most of whom also serve on the boards of Washington state post-secondary institutions, to give him guidance on the educational needs into the second decade of the new millennium.

3. T. Marchese, "Not-So-Distant Competitors: How New Providers Are Remaking the Post-Secondary Marketplace," American Association for Higher Education bulletin, May 1998

4. Cited in the Christian Science Monitor, June 30, 1998, p. B7.

5. I am the principal investigator of a national study of distance education in political science. Montgomery Van Wart, Peter Dombrowski, and Mack Shelley are also on the research and assessment team.

6. S. D. Keene and J. S. Cary, "Effectiveness of Distance Education Approach to U.S. Army Reserve Component Training," American Journal of Distance Education, # 4, 1990, pp. 14-20.

7. J. Wergin and J. W. Haas, "Televising Graduate Engineering Courses; Results of an Instructional Experiment," Engineering Education, November 1986, pp. 109-112.

8. T. J. Delonghy, "Remote Instruction Using Computers Found as Effective as Classroom Sessions," The Chronicle of Higher Education, April 20, 1988, pp. A15, A21.

9. Las Cruces: New Mexico State University 1997.

10. This information comes via e-mail from Mike Albright, ISU, on a POD discussion of distance education, June 1998.

11. I have experimented for several years with "remote participation" techniques. I've had excellent results by my own assessment. Students are very satisfied with the quality and results, according to careful evaluations in which they've participated.

12. We have also added CD-ROMs, a rich supplement of electronic and other teaching tools, and now a new, powerful World Wide Web support system that can be browsed at http://www.schmidt.politics.wadsworth.com.

CHAPTER 7

Asking the Right Questions: A Five-Step Procedure for Incorporating Internet Technology into a Course

Siaw-Peng Wan
Asst. Prof. of Business Administration

Elmhurst College
Elmhurst, IL

INTRODUCTION

Since the establishment of the Internet in 1969, educators have been incorporating its use in one form or another into their courses. However, such uses were not widespread because access to the Internet was limited to a number of higher-learning institutions. In addition, the Internet was a static, text-based, non-user-friendly, and non-interactive environment. This situation changed in 1992 when the World Wide Web (commonly known as the Web) was established as the newest component of the Internet.

Unlike the other components of the Internet, the Web is capable of delivering multimedia components (e.g., graphics, sound, video, etc.), which provide a visually richer environment. The navigation process has also been simplified with hyperlinks being embedded in a Web page that allow users to "jump" to a different Web page simply by clicking on the hyperlink, instead of needing to know arcane commands. With the commercial control of the Internet backbone in the mid-1990s, Internet accessibility increased dramatically when a large number of Internet service providers began offering access to the public. The commercialization of the Internet has also fueled explosive growth in new Internet technology and tools, which further enhanced its multimedia and interactive capabilities.

These new developments in Internet technology have created a high level of excitement among many educators because it became easier for them to incorporate use of the Internet into their courses. As Internet technology matures, it is beginning to find its way into many educators' courses. However, there are still a large number of educators who are interested in incorporating the Internet into their courses but have no idea where to begin.

This chapter will present a five-step procedure that prepares educators before they attempt to incorporate Internet technology into their courses for the first time. This five-step procedure will guide educators through the planning stage of incorporating the Internet into their courses.[1] Each of the five steps represents a question that will guide educators through the process.

Step 1: "Why do I want to use the Internet in my courses?" helps educators establish the goals they would like to accomplish with the technology.

Step 2: "What are various tools and supports available to me?" helps educators identify all the available resources.

Step 3: "What are some of the problems I might encounter?" helps educators understand some of the potential obstacles they might face.

Step 4: "What are the different ways I can use the Internet to achieve my goals?" helps educators model the various Internet components they would use in their courses.

Step 5: "When should I complete all the different Internet components?" helps educators set up the initial timeline for completion.

This chapter is structured as follows: The next five sections will discuss some of the issues associated with each of the five steps and provide examples and ideas that will help educators begin the process; the final section will provide some concluding remarks.

ESTABLISHING GOALS

The first and perhaps most important step of the five-step procedure is to establish goals that an educator would like to accomplish in a course with the aid of Internet technology. Many educators make the mistake of incorporating the Internet into their courses without first identifying what they would like to achieve with the technology. They are using the technology because everyone else is using it, instead of trying to change, improve, or supplement what they are already doing in their courses.

Asking questions such as, "Why do I want to incorporate the Internet into my courses?" or, "What would I like to accomplish in the course with the aid of the Internet?" will help an educator begin the process of establishing goals. At this stage, it is not important for educators to develop a list of goals that are feasible. It is more important to brainstorm at this stage and worry about the feasibility of realizing some of these goals at a later stage.

Different educators will have different goals; some of them will have simple goals, while others will have more elaborate ones. Generally, these goals can be grouped into four categories: (a) improve information exchange; (b) use new resources; (c) develop skills with the aid of Internet technology; and (d) create an alternative and/or supplemental learning environment.[2]

Improve Information Exchange

The Internet is a medium that can be accessed anytime and anywhere. With the aid of this new medium, educators can improve information exchanges with their students. There are two ways to achieve this goal: Provide a means for students to access course information with ease, and improve communication between educators and students. To provide easier access to course information, educators can set up information centers for their courses on the Internet (more specifically, the Web). Such an information center represents the gateway to all course announcements (e.g., quiz schedule, assignment deadlines, etc.) and relevant materials (e.g., syllabus, assignments, lecture notes, etc.) Basically, the online information center has replaced the reserve section of the library.

The main advantage to educators is that they do not need to deal with the logistics of individually contacting students when there are changes to the courses. All they need to do is update the information on the course information center. As for the students, the advantages for having such an online information center are that they always have up-to-date course information, they have access to the information anytime and anywhere, and they no longer have to contend with unavailable or missing materials.

The Internet also provides the means to communicate electronically, which allows students to communicate with educators with greater ease and flexibility. Students are no longer limited to meeting with educators during scheduled office hours. It is possible for them to contact educators at their own convenience and for the educators to reply at their own convenience. Students can seek feedback on questions and assignments without being physically present on campus. Such opportunities are especially important for commuter students who are not available during regular office hours due to other commitments.

Use of New Resources

As the Internet (particularly the Web) matures, an increasing number of resources, such as data, articles (from journals, popular magazines, daily newspapers), up-to-date news item, etc., are available to the general public. The Web has presented new sources of information to students and educators in addition to those found in the library. It is important for students to take advantage of the wealth of information on the Web for their assignments and projects because many of these sources were previously unavailable or difficult to access, and they provide more current information than those available in the library. Such resources also play a valuable part in helping educators keep the materials that are covered in class current, relevant, and exciting.[3]

Develop Skills with the Aid of Internet Technology

Incorporating the Internet into a course will help students get more comfortable with using the technology. However, Internet technology can also be used to help students develop or improve valuable skills, such as communication, research, and collaborative skills.

Communication Skills

Electronic communication can help different groups of students in a class develop different levels of communication skills. It allows students who are shy or reluctant to participate in class, and foreign students who are uncomfortable with the language, to have the opportunity to participate in an online group discussion. Other more vocal students will not intimidate them because they have ample time to compose their questions and/or responses. As for the more vocal students, electronic communication encourages them to put more thought into what they have to say because they have to type out the responses rather than saying whatever comes to mind, and they know that all their comments are archived for future reference.

It is important to point out that online discussion should serve as a supplement to, and not a replacement for, in-class, face-to-face discussion. Proper use of online discussion will lead to improved in-class discussion by helping students who are shy or uncomfortable with the language build up confidence in their communication skills, and by helping the more vocal students become accustomed to putting more thought into their responses.

Research Skills

As the amount of information available on the Internet becomes more and more abundant, it is important for students to develop better research skills to search for and to evaluate the information acquired. It is crucial for them to develop better search skills to seek out information in an environment where the resources are categorized in a structured manner. Once they have found the information, they need to have the skills to evaluate the relevance and validity of the sources. Unlike printed resources, many of the Internet resources are non-reviewed, non-researched, and inaccurate because anyone with Internet access can easily publish anything on the Internet.

Collaborative Skills

It is common for educators to break up their classes into smaller groups in which the students are required to collaborate with others to complete an assignment or a project. Such collaborative skills are important for students later because they will be required to work closely with their colleagues and clients. With the Internet, it is no longer necessary for individuals to be physically present to collaborate on a project. In addition, there is also a recent flood of software packages that improve collaboration over the Internet. As a result, electronic collaboration is increasingly becoming the norm in the real world. Hence, it is important for students to be comfortable with and to understand the proper etiquette of conducting such online collaborations.

Create an Alternative and/or Supplemental Learning Environment

The accessibility, flexibility, and interactive and multimedia capabilities of the Web have provided educators a means to create a learning environment that supplements or provides an alternative to the traditional classroom environment. Owston (1997) indicates that the Web presents a new way of promoting improved learning among students for three main reasons: It appeals to the current generation of students' learning mode; provides for flexible learning; and enables new kinds of learning. We will look at examples of some alternative and/or supplemental learning environments that are feasible with the aid of Internet technology.

By incorporating the Internet into their courses, educators can help develop an alternative learning environment that focuses more on active learning rather than passive listening. Educators can reduce their number of lectures by providing relevant materials to the students in advance and focusing more on discussion in class. Internet technology can also be incorporated in other ways to help students focus on how they can apply the materials learned in the course and observe how those materials are applied in the real world. In addition, the Internet can also be used to provide experiences not easily achievable or possible in a traditional classroom. For example, in a virtual setting, students can visit the Louvre in Paris or participate in an archeological dig in Egypt.

Educators can also choose to use Internet technology to supplement their classrooms rather than modify it. For example, educators can take advantage of Web-based materials not covered in class to help pique the students' interest and broaden their knowledge in a particular area. Educators can also provide Web-based interactive learning components that will help students review some of the materials covered in class. These components can be designed to provide appropriate feedback automatically when necessary. In addition, educators can also provide avenues where students can explore certain subjects on their own outside the classroom. With the multimedia and interactive nature of the Web, it is possible for educators to design components that provide the students the opportunity to visualize and/or experiment with certain concepts on their own.

IDENTIFYING AVAILABLE RESOURCES

In the previous section, we presented some of the goals educators might like to accomplish in their courses with Internet technology. However, it is not necessary for them to attempt to accomplish every one of them. Educators should choose only those that are most appropriate for their courses. Once they have established the goals they would like to accomplish, the next step is to conduct an inventory checklist of available resources and potential future resources on campus. These are the resources that will help educators realize the goals they have established. Identifying the resources at this stage will ease some of the frustration and anxiety when the time arrives to actually incorporate the Internet into the courses. Educators need to pay attention to three areas of resources: financial; technological; and educational.

Financial Resources

The expenses associated with incorporating Internet technology into a course depend greatly on the types of technology used. Educators who are attempting this for the first time should always start out with something simple (e.g., using e-mail for communication). In many instances, the tools are already available or can be acquired with minimal expenses. However, when educators are ready to venture into more advanced Internet technology that provides multimedia and interactive capabilities, it will be necessary to purchase certain expensive software packages. In addition, educators will also need funding to attend conferences and workshops that focus on Internet-related pedagogical issues and techniques. These are generally excellent avenues for educators to share ideas with others and/or to receive tips and ideas from Internet-experienced educators.

It will be helpful when the time arrives to purchase the software and/or to pay the conference registration fees, for educators to know what are the available financial resources on campus that will fund their software acquisitions and/or conference attendance. The first place to look for financial support is to determine if the departments of the educators are willing to support them in using Internet technology in their courses. Other financial resources at an institution could include curriculum development or technology grants that encourage innovative pedagogical techniques in the classroom.

Technology Resources

The two areas of technology resources that educators need to be aware of are computer resources and Internet technology expertise. When it comes to incorporating the Internet into the courses, computer resources represent the software and hardware available on campus for educators, whereas Internet technology expertise represents the source of help and guidance for educators.

The best way for educators to determine the available computer resources on campus is to approach the information system or computer service department. This department generally will be able to provide crucial information regarding available Internet-related resources, such as software for creating Web pages, types of Web browsers, software for running newsgroup and listserv, etc. It is important for educators attempting to incorporate the Internet into their courses for the first time not to deviate from what is available because computer service departments generally do not provide assistance for resources not supported.

Seeking out the sources of Internet technology expertise on campus is a crucial step at this stage. It is important for educators to seek the advice of these sources before they begin the actual process of incorporating the Internet into their courses. This is because they can generally provide tips that will ease the tasks and point out pitfalls that should be avoided.

There are three potential sources of Internet technology expertise available at an institution. The first is the information systems or computer service department. Depending on the size of the institution, the assistance these departments can provide for educators ranges from setting up a simple course Web site to programming complicated Internet components. The second source is other educators on campus who are currently using the Internet in their courses. They are usually a very good source because they can provide first-hand information regarding the successes and failures of using certain types of Internet technology for pedagogical purposes. The third source of Internet technology expertise is the technology and curriculum office or group on campus. The purpose of this office or group is to assist educators in using technology in their courses. Ask if they will provide assistance for using Internet technology. It is important for educators to understand that it is not necessary that all three resources be available to them.

Educational Resources

The best way for educators to get started using the Internet in their courses is to attend workshops on campus designed to introduce the various types of Internet technology and how they can be used in a course. These workshops usually generate a lot of ideas for educators on how to use various types of Internet technology to best achieve the goals they have established.

They are usually conducted by the technology and curriculum office or by other educators on campus. However, if they are not available on campus, educators should be alert for similar workshops at other institutions or organizations that are open to the public. These workshops are generally free or reasonably priced. In addition, there are also conferences that focus solely on Internet-related pedagogical issues and techniques. However, these conferences are moderately expensive, in general.

UNDERSTANDING THE OBSTACLES

It is important for educators to identify available resources on campus that will help them realize the goals they have established for using the Internet in their courses. However, it is also crucial for them to understand that the lack of such resources may become obstacles that will affect the feasibility of realizing some of those goals. There are also other potential obstacles that educators might face. The three most common are time constraints, unavailability of off-campus Internet technology, and lack of advanced computer support.

The biggest obstacle faced by educators is finding the time required to incorporate Internet technology into their courses effectively. It is important for educators to understand that incorporating Internet technology into a course does not lighten the educator's traditional job responsibilities (i.e., teaching, research, and service). In reality, for many educators it actually means devoting additional time for searching and evaluating Web resources, reading and responding to e-mail, creating or modifying Web pages, developing Web-based components, etc. There is an even greater time commitment for educators incorporating the Internet into their courses for the first time because they need to spend the time learning and mastering this new technology. Since time is a very valuable commodity for most educators[4], they will be limited to using components that do not take extensive time to create. This means that the educators might not be able to effectively achieve all their established goals.

Student access to off-campus Internet technology represents another obstacle for educators with a large number of commuter students in their classes. Educators need to ask questions such as, "Do they have Internet access?" and, "Do they have the appropriate software?" It is pointless to create an online information center or other Internet components when the students do not have access to the Internet or the appropriate software to view the components properly. Hence, the availability of off-campus Internet technology to commuter students will have a direct impact on how effectively educators can incorporate the Internet into their courses.

The third obstacle faced by educators is the availability of advanced computer support. This obstacle is not as severe as the previous two because educators can still create many Internet components without the availability of such support. However, they will be limited in what they can create because Internet components that provide certain levels of interactivity require knowledge of advanced computer-programming languages. Since most educators do not have an extensive background in programming, or the time and resources to learn it, they will be limited in how they can effectively create Internet components that will achieve their goals.

MODELING THE COMPONENTS

Once the educators have established the goals, identified the available resources, and understood the potential obstacles, it is time for them to model the various components that will help achieve the established goals with the given resources and obstacles.[5] It is important for educators to understand that there are many ways to model Internet components to meet the established goals. This section will provide some examples (or ideas) on how the Internet can be used to realize some of the goals outlined earlier.[6]

Improve Information Exchange

The easiest way to provide an information center for a course is to set up a Web site for that course. This is where the students can access all relevant course information and materials. Educators can post announcements regarding their courses, such as changes or errors in assignments, upcoming quizzes and exams, time and location of review sessions, etc. In addition, students can also access course materials such as syllabi, assignments, lecture materials, old quizzes and exams, etc. on the Web site.[7] However, the Web site does not need to be limited to providing course information and materials to the students, it can also serve as the gateway to other materials, such as links to Web resources, online discussion forums, Web-based interactive learning components, etc.

To encourage students to use e-mail as a communication tool, educators need to provide incentives for them to do so. One incentive would be to award some bonus points for written assignments that are turned in electronically. Another would be to allow students to seek feedback on their papers or parts of their projects by sending them as attachments in e-mail. The educators can then provide the feedback directly on the documents and send them back to the students.[8]

Use of New Resources

There are many ways educators can encourage students to take advantage of Internet resources. Providing links to relevant Internet resources on the course Web site is the easiest way for educators to introduce students to resources available on the Internet. However, it is important for educators to research and evaluate the resources before linking to them on the course Web site. Educators can also require students to search for other relevant Internet resources as an assignment. This is a good way to build up Internet resources for future classes.

Another way to encourage students to take advantage of Internet resources is to have specific assignments for which students are required to search the Internet for information and/or data. These assignments can be designed to provide different levels of assistance in using the Internet. Educators will provide links to relevant resources for the first few assignments but provide only hints on where to look for the resources for later ones.

A third way to encourage students to take advantage of Internet resources is to allow them to use such resources on their papers and projects, with the stipulation that they also have to use printed resources from the library. It is important that educators encourage students to use both Internet and printed resources, and not rely only on Internet resources.

Develop Skills with the Aid of Internet Technology

Earlier, we discussed the possibilities of developing or improving a student's communication, research, and collaborative skills with the aid of Internet technology. In this section, we will look at a few examples of how this can be accomplished.

One way for educators to improve or develop students' communication skills is to provide an online forum for them to discuss a specific topic.[9] To make sure that every student participates in the discussion, each will be required to contribute to the forum a predetermined number of times within a specific time period. It is important that educators monitor such discussions so that they do not get out of control. Certain rules need to be applied regarding the etiquette of online discussion.

To improve students' research skills, educators can give assignments in which students are required to search the Internet for relevant information. In addition to submitting the answers to the assignments, students will be required to turn in a short write-up of the process they used to search for the information needed and discuss the validity of the resources. It will also be beneficial if educators have students share their experiences and search techniques when they go over the assignments.

Since one of the strong points of the Internet is ease of communication, regardless of geographical boundaries, educators can take advantage of this to help their students develop better collaborative skills. Arrangements can be made with educators at other institutions so students from different parts of the country and/or different parts of the world can work on projects over the Internet. Each group will contain students from different institutions who will collaborate in their assigned online forum. The projects can be designed for students from similar courses, for students from different courses in the same discipline, or for students from courses in different disciplines. For example, educators from the United States and South Africa can set up a comparative study to look at the human rights issues in those two countries. In addition to discussing (or debating) issues, students will also learn to work collaboratively with other students who have completely different cultural backgrounds.

Create an Alternative and/or Supplemental Learning Environment

The continued advancement in Internet technology, since the introduction of the Web in 1992, has made the Internet a medium that can help transform a traditional classroom environment into an active learning environment. Teaching and learning is no longer limited to the classroom, but is conducted outside the classroom as well. Advanced Internet technology enables educators to provide an alternative and/or supplemental learning environment by creating interactive learning components that encourage students to be active learners.[10] We will look at several examples on how educators can use Internet technology to provide a different learning environment, which cannot be easily replicated in a traditional classroom.

As discussed earlier, an online discussion forum can be used to improve students' communication skills and to provide a means for collaborating in projects. Educators can also use online discussion forums to create an active learning environment for their students. They can accomplish this by inviting experts in the discipline to participate in the discussion.[11] The experts can help students understand how materials learned in class apply in a real-world environment. Different perspectives can be shared on how certain techniques or approaches need to be modified to accomplish given tasks. The experts can also enhance the active learning process by posting actual scenarios and having students discuss what the appropriate approaches should be. After the initial stages of online discussion among students, the experts can share with the students the approaches they have taken. Further discussions are possible to explore the merits of the experts' approaches and the possibilities of other alternatives.

Educators can also use advance Internet technology, such as scripting language (e.g., JavaScript) or a programming language (e.g., Java) to create components that help students review course materials more effectively. For example, educators can use either a scripting or a programming language to create online quizzes, with the goal not being to score the students but to give them the opportunity to review the course materials on their own. Educators can achieve this by creating online quizzes that provide tailored feedback based on students' choices. When a student chooses the wrong answer, feedback will be provided on why the choice is wrong and hints will be provided to lead the student back on the right track. Even if the student picks the right answer, feedback will still be provided on why the answer is right. This is important because the student might have guessed the answer without really knowing why it is correct.[12]

It is not necessary for educators to have extensive knowledge of computer programming to create interactive learning modules that help students become better learners. There is a large number of authoring software packages that allow educators to create interactive modules without extensive programming background. The goal of these modules is to allow students to focus on learning the concepts. In certain disciplines, they allow the students to explore and experiment with the concepts on their own without being bogged down by solving the models or performing numerical calculations. Most of the time, students are so busy with the equations or the calculations that they completely forget what they are supposed to learn.[13]

One such example is the concept of the relationship between the shape of an efficient frontier and the correlation between the returns of two financial assets in finance. Students often need to perform numerous calculations in order to plot an efficient frontier and, hence, it is often not feasible to have them look at various scenarios in a classroom environment. Most of the time they are so bogged down by the calculations that they are not paying attention to how the shape of the efficient frontier changes as the correlation changes. Interactive learning modules can be created to give the students a visual representation of the shape of the efficient frontier (without performing any calculations) as they change the correlation.[14] Such modules can make learning fun, and also allow the students the freedom to explore different possible scenarios.

The types of interactive components created depend on the instructor's programming experience and/or computer support. Some of these components require a great amount of knowledge in programming. However, it is not necessary for educators to create all the components themselves for their courses. It is possible to collaborate with other educators to develop the components jointly, or they can use components created by other educators (at the same or a different institution), provided that permission has been granted.

Setting Up the Initial Timeline for Completion

When educators come to the final step of the five-step procedure, they will need to set up an initial timeline for creating the various components they have modeled to achieve their established goals. The timeline serves as a guide and does not need to be very detailed in terms of exact dates for beginning and completing each component, but it should indicate which components will be created in each semester.

There are several factors educators need to consider when they are setting up the timeline. They need to understand that it does take time to learn and master the technology needed to create some of the components. In addition, they need to understand that it can be difficult to find time to create the components when they have other responsibilities, such as family, teaching, research, committee work, etc. As a result, educators should not be too ambitious the first time they introduce Internet technology into their courses. It is not necessary for them to have all the components available the very first time. It might potentially take several semesters before all the components begin to take shape.[15] Hence, it is best that educators create the components one piece at a time.

Educators should always start by providing existing materials (such as syllabus, regular assignments, lecture notes, etc.) on the Web site, then progress to materials that can be easily incorporated, such as links to Web resources and assignments using those resources. Reserve the more difficult components, especially those that require programming skills or expensive software, for a later date when such resources become available. It is always easier to obtain financial support on software purchases when educators can show administrators how the various software packages are needed to further what they are trying to do.

The timeline will help educators monitor when each of the components will be available for use by students in a particular semester. This is very important because educators should design their courses for each semester based on the availability of the various components. Failure to do so will result in frequent changes in the course design because some of the components needed are not ready. In addition, they should always have the components they want to use in a particular semester ready prior to the beginning of that semester. For educators to attempt to create the components during the semester in which they are needed can be disastrous because it is very easy for them to get busy with something else and fail to have the components completed. Having the components completed before the semester will help prevent the situation of over-promise and under-delivery.

CONCLUSION

The development of the Web as the newest component of the Internet has provided educators a tool to improve or enhance what they are already doing in the classroom. The ease of accessing information on the Internet enables educators to easily disseminate information to the students. And the wealth of information available on the Internet provides educators and students with new resources. In addition, the multimedia and interactive capabilities of the Internet enable educators to help students develop or improve some of their skills and provide an alternative and/or supplemental learning environment. As a result, there has been a growing trend among educators in using the Internet in their courses. However, there are still a large number of educators who are interested in incorporating the Internet into their courses but find this a daunting task because they have no idea where to begin.

The chapter has outlined a five-step procedure that will guide Internet-inexperienced educators through the process of incorporating the Internet into their courses. It gives a series of five questions to help educators establish the goals they would like to accomplish with the technology, identify available resources, understand potential obstacles, model various Internet components to achieve the goals, and set up the initial timeline for completing the various components. The procedure represents the first step (i.e., the planning stage) educators can take towards using Internet technology in their courses.

Once the educators have established a blueprint for incorporating technology into their courses, they are on their way to the second stage, which is the creation of the various Internet components they have modeled. However, it is still crucial for educators to revisit the five-step procedure whenever they encounter changes that affect any of the five steps. For example, the goals might have changed, additional resources may become available and/or existing resources may have disappeared, existing obstacles may have been overcome and/or unanticipated obstacles may have been encountered, or they may have new ideas for the components and/or found that some of the components not feasible.

References

Chickering, A.C. and Ehrmann, S.C. "Implementing the Seven Principles: Technology as Lever." August 28, 1997. http://www.aahe.org/technology/ehrmann.htm. (February 16, 1998).

McCormak, C. and Jones, D. Building a Web-Based Education System. New York, NY: John Wiley & Sons, Incorporated; 1998.

Owston, R.D. "The World Wide Web: A Technology to Enhance Teaching and Learning?" Educational Researcher, Vol. 26, No. 2; March 1997, pp. 27-33.

Porter, Lynnette R. Creating the Virtual Classroom: Distance Learning with the Internet. New York, NY: John Wiley & Sons, Incorporated; 1997.

Serim, F. and Koch, M. Net Learning: Why Teachers Use the Internet. Sebastopol, CA: Songline Studios, Incorporated and O'Reilly & Associates, Incorporated; 1996.

Walsted, W.B., Fender, A.H., Fletcher, J., and Edwards, W. "Using Technology for Teaching Economics," Teaching Undergraduate Economics: A Handbook for Instructors. New York, NY: Irwin/McGraw Hill, 1998, pp. 269-285.

Endnotes

1. Educators generally go through four stages when they incorporate the Internet into their courses. These four stages are planning, production, implementation, and assessment. This chapter focuses only on the planning stage.

2. These categories were developed through the author's own experience and numerous discussions with other educators at workshops and seminars. This is not an exhaustive list of all the possible goals that an educator might like to accomplish. It is here to help some educators get started with the process.

3. Publishers are beginning to provide companion Web sites for their textbooks to educators who adopt them. These Web sites generally provide reviewed and updated resources that are relevant to the textbooks.

4. Educators at a large number of institutions face the situation where a much greater emphasis is placed on scholarship development rather than pedagogical development. Those educators are under great pressure to publish for tenure and promotion. As a result, most of

their time will be devoted to their research and not to pedagogical development, such as the use of the Internet in their courses

5. This is the stage where educators, based on the available resources and obstacles, generate ideas on the various ways Internet technology can be used to achieve the established goals. However, this is not the stage where they actually create the components. That will be done during the production stage.

6. There are many more specific examples on how the Internet can be incorporated into a course. The examples listed here are some of the ones the author and his colleagues have used in their courses. The following are some excellent resources for additional examples on how the Internet can be used in a course: Serim and Koch (1996), Porter (1997), Chickering and Ehrmann (1997), McCormak and Jones (1998), and Walsted, Fender, Fletcher, and Edwards (1998).

7. There are three ways educators can provide these materials on the Internet: (a) create them in hypertext markup language (HTML) from scratch; (b) convert existing materials in their native formats into HTML, or (c) provide links to existing materials in their native formats. The first two ways allow the students to view the materials with the browser, whereas the third requires the students to download the files from the server and view them with the appropriate software. All three ways can be easily accomplished with a Web page editor such as Microsoft FrontPage or Adobe PageMill. The advantage of using a Web page editor is that educators do not need to have extensive knowledge of the HTML that generates the Web pages. In addition, educators can also use course management software packages, such as Web Course in a Box, which are designed specifically to help them create course Web sites.

8. The feedback should be typed in a different style or font to distinguish it in the documents.

9. The online forum can be conducted with a Web-based bulletin board system, newsgroup, listserv, or a course management software package (e.g. WebCT).

10. Examples of advanced Internet technology include the Netscape plug-in and Microsoft ActiveX architectures, programming languages (e.g., Java), and scripting languages (e.g., JavaScript and VB Script).

11. This can be accomplished with a bulletin board system for asynchronous communication or with a chat room for synchronous communications. Each medium has its advantages and disadvantages. A bulletin board system provides students and experts more time to compose their comments and it also archives all discussions for further reference. However, there is no spontaneity to the discussions. On the other hand, chat room allows real-time communication among students and experts for a more spontaneous discussion. However, there is less time for composing a good response.

12. An example of an online quiz that provides instant feedback based on a student's choices can be found at Elmhurst College (www.elmhurst.edu/~cbe/vcroom). This quiz is created with JavaScript and is limited in what it can do. If educators need a more powerful and flexible online quiz, it will need to be created with a programming language such as Java. There are also software packages, such as Asymetrix Toolbook, which enable educators to create similar quizzes without having any programming knowledge.

13. It is important to point out that we are not downplay the need for students to know how to do those calculations. These modules are designed to help students focus on understanding the concepts first without being distracted by the calculations software packages, such

as Asymetrix Toolbook. That enables educators to create similar quizzes without having any programming knowledge.

14. An example of an interactive learning module that teaches the relationship between the shape of an efficient frontier and the correlation between the returns of two financial assets can be found at Elmhurst College (www.elmhurst.edu/~cbe/vcroom). This particular learning module is created with Formula One Net Pro. Other interactive learning modules can also be created using software such as Macromedia Director and ASAP Web Show. An excellent source for educators to generate ideas for various interactive learning modules is to visit Netscape's plug-in page (home.netscape.com/comprod/products/navigator/version_2.0/plugins/index.html). Each of the software packages listed here provides a demonstration of how it can be used for various languages, such as Java. An example of an interactive learning module created with Java is the FlyLab at California State University (www.cdl.edu/FlyLab), which teaches students the principles of genetic inheritance with virtual mating of fruit flies.

15. It is important for educators to understand that the initial timeline serves only as a guide and that it should be modified when there are changes in available resources, the educator's schedule, etc.

CHAPTER 8

When Less Is More: Some Ergonomic Considerations in Course Page Design

Suba Subbaro

Instructor of English

Oakland Community College

Auburn Hills, MI

Laura Langa-Spencer

Graphic Designer

Let's Get Graphic

Birmingham, MI

INTRODUCTION

Instructors new to online teaching are usually most concerned with putting as much information as they can into their course pages. Frequently, they also feel bound to use all of the decorative features offered by the Web publishing software they are using. Moreover, they tend to be overly optimistic about their students' experience with computers and assume that those who can surf the Net also know how to learn in this medium.

Students, on the other hand, especially if they are new to online learning, grapple with vastly different issues. They are overwhelmed not just by the amount of information available to them but also by the constant need to make decisions about which links they should follow. They skim the pages impatiently, follow links impulsively, and end up more often than not "lost in hyperspace." They simply print out every page their instructor requires them to read. And after the first visit or so, they may not even notice the spinning GIFs and scrolling marquees that their instructor has taken so much trouble to include.

It is in the context of these conflicting expectations and choices that this chapter offers its pragmatic recommendations for a user-oriented design of course pages.

PRELIMINARY CONSIDERATIONS

Student Behaviors That Affect Course Page Design

Some specific behaviors of on-line learners, as well as cognitive issues relating to online learning, deserve consideration at this point.

The Lure of Hard Copy

Many online instructors are surprised that their students print out so much information. They believe that this is entirely unnecessary and uneconomical.[1] From the students' perspective, though, this is simply practical—after all, even the slimmest laptop computer cannot beat the easy readability and portability of the printed page. Since online learning is still not the norm, it is reasonable to believe that many students in the average online class will still seek the tangible, tactile comforts offered by the printed page. Therefore, the likelihood that course pages will be printed cannot be ignored by instructors/designers.

Cognitive Foundations of Reading

Since hypertext is, after all, a kind of text, and since online students have to depend largely on the written word, a brief consideration of how people read can also offer course page designers valuable insights. Most cognitive models of reading posit that successful readers read by creating in their minds a meaningful, well-structured representation of the text and integrating what they have just read with what they already know (Charney 243-8). Consequently, these tasks are facilitated if important ideas are presented early in the text, with the connections between them made explicit and the overall presentational structure revealed and made familiar to the readers. If these cognitive expectations are ignored, hypertext, far from being the liberating medium it has been hailed to be, can easily become confusing and constricting.

Cognitive Challenges Presented by Hypertext

A third important consideration is that students don't always come to an online class knowing how to learn in cyberspace. Hypertext, for all its virtues, places a lot of pressure on students in terms of information overload and what Conklin terms "cognitive overhead," by forcing them to constantly choose where to go next or to remember where they have just been or simply to understand where they are (Conklin 40). Most newcomers to online learning tend to bring to hypertext the same linear approaches they use to read print text and find the absence of the traditional structural cues disorienting. They might simply surf the material presented and come away with irrelevant or incomplete information. Therefore, in view of how students read and how they read online text in particular, instructors must create course pages that both disseminate information *and* guide students in reading this information efficiently.

Environmental Variables That Affect Course Page Design

The technological environment in which online students work must also be taken into account in the design of efficient course pages.

Hardware and Software Differences

Factors such as equipment capabilities, the amount of memory available, and even the willingness to use plug-ins can all affect the ease and comfort with which students learn online. While it would be impossible for designers to accommodate all the available options, they must be aware of how different monitors, platforms, and browsers can cause online text to be received differently.[2] For instance, graphics tend to look darker in Windows due to higher gamma settings but are relatively much lighter on a Macintosh; a 12-point typeface is much smaller on the Mac than on Windows; and both Netscape and Microsoft Internet Explorer, the two most popular browsers today, have each created specific HTML tags that will work only in their browsers.

Customizable Options

Even if it can be assumed that all students in a class are using comparable equipment, the ability of individual users to customize options must still be considered. Graphics, for instance, might be turned off in browsers in order to increase download speed. Or, the course page designer's choice of text and link colors and even font can be overridden by the savvy user who still might decide to print out the pages on a black-and-white printer. These possibilities must be addressed and accommodated, at least to a certain extent, if Web pages are to preserve their essential meaning under most circumstances. Thus, even with student groups with no special access issues, instructors must still consider how the variables described above will affect the design of their course pages.

Guidelines for Ergonomic Course Page Design

The following recommendations first address the general needs of all Web users and then the specific needs of online learners.

Considering General Usability Issues in Web Page Design

Optimizing Download Times

The main problem for Web users, frequently observed and well documented in surveys, is speed. A recent study by the Graphics, Visualization, and Usability (GVU) Center at Georgia Tech, for example, revealed that 63% of users report this problem (8th User Survey). While this percentage is lower than that reported in earlier surveys, it is still significant. Consideration of the time it takes for a page to download is, therefore, of fundamental importance to Web page designers because it relates to a number of other design issues as well. Research into response time values indicates that Web users can tolerate only about 10-15 seconds of Net lag (Nielsen "The Need for Speed"). If the system response time is longer than that, users tend to move on to other choices.

A number of variables can affect download times: the server where the site resides; the server's connection speed; activity on the Internet; the user's connection speed; the rendering speed of the user's browser; and the capabilities of the user's hardware, among other things. Therefore, it is imperative to remain conservative about file size and check download times for every page. A 30K file, for instance, takes 20.9 seconds to download on a 14.4 kbs modem; 11.54 seconds on 28.8kbs; 7.78 seconds on 56kbs; 2.81 seconds with an ISDN connection; and 1.1 seconds using a T1 line. Designing for files to be downloaded on a 14.4kbs modem would be a safe choice.

Download times can be checked quite easily using the utilities that come with some of the text editors or by visiting a site such as the Web site Garage (http://www.websitegarage.com). More importantly, following some simple strategies can optimize download times. Images should be used only if they fall into one or more of the following categories: indispensable; text-supportive; or clearly symbolic. If their purpose is merely cosmetic, they do nothing but stress user bandwidth and detract from the readability of the site.[3]

When images must be used, creative cropping can reduce image file size without distortion. Using the right graphic file format can also make a difference.

Generally speaking, photographs are best saved as JPEG files, while images such as cartoons, logos, or illustrations with large areas of solid color can be better presented as GIF files. Download speed can also be improved by repeating key design elements throughout a set of pages—this cannot only create a sense of "brand identity" but also allow for rapid retrieval of images from the cache instead of necessitating a separate connection to the server for each graphic file. It is also a good idea to alert users by indicating file format and size whenever a page might stress user bandwidth.

Using Frames, Graphics, Animation, and Multimedia

Newcomers to Web page design tend to overlook the fact that every frame, image, sound file, or video clip included on a page requires a separate connection to the server. Naturally, these repeated connections encroach upon the 10-15 seconds that the average user is willing to expend in waiting for the whole page to download. Moreover, any movement on the page—an animated GIF or a scrolling marquee, for instance—places an additional cognitive burden upon the reader by engaging the peripheral vision and thereby distracting the attention from the very text which the animation was intended to highlight (Nielsen "Top Ten").

Every addition to the page needs to be ruthlessly scrutinized for the value it adds to the page. Images, in particular, may not always be necessary. If it is merely an enhancement, care must be taken to ensure that it is the important *textual* information that downloads first and appears at the top of the page. For instance, the setting of height and width attributes for each graphic will cause browsers to load the text first and the graphic later. A thumbnail version of a large image, with a parenthetical alert about the file size, also allows the readers the option of selecting the entire image or of bypassing it altogether. Meaningful <ALT> coding of all images ensures that even when the loading of graphics is turned off in the browser, the reader still gets the essential information. Figures 8.1-8.3 illustrate the importance of <ALT> coding of images. It can be easily seen that in Figure 8.3 (on page 115), essential information is actually lost if the user decides to turn off auto loading of images in the browser.

Text loading can never be turned off! Designers would do well to take advantage of this fact. Even background graphics can be a distraction. For one thing, they can make the text less legible than otherwise, even if they are lightly textured. For another, different browsers (and even different versions of the same browser!) can display colors differently. So, as there is no guarantee that what the designer sees is what the user will get, it is best to limit oneself to using solid background colors that contrast strongly with text colors.

Figure 8.1 *A Page with Images*

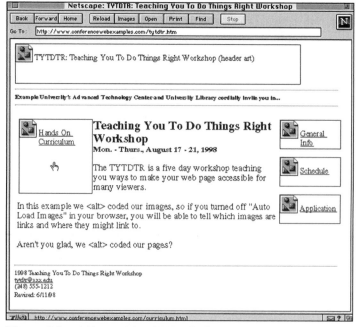

Figure 8.2 *A Page with <ALT> Coding of Images*

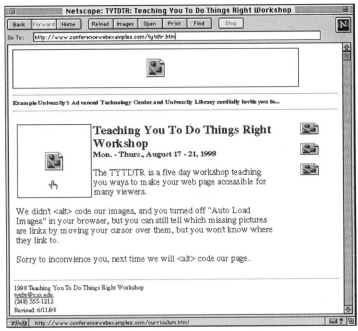

Figure 8.3 *A Page Without <ALT> Coding of Images*

Surveys conducted by the GVU indicate that the majority of users (46%) prefer black text on white background (8th User Survey). While white on black is also acceptable, it would be a hindrance to the reader who decides to print the page because the general preferences on the browser will have to be changed first if the page is to be readable in hard copy. The white-on-black design in Figure 8.4 might

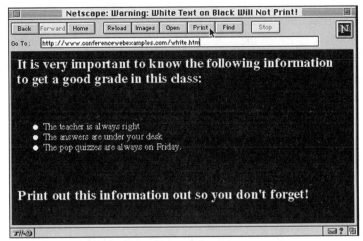

Figure 8.4 *Why Black Backgrounds May Not Work*

look striking on the screen, but it will show up as a blank page if the user decides to print it.

Likewise, proprietary tags such as <MARQUEE> in Microsoft Internet Explorer or <SPACER> in Netscape, and tools such as Java and Adobe Acrobat PDF files (or anything that requires a plug-in) should be considered enhancements, not the primary means of conveying essential information. It is, therefore, a good idea to use a basic version of HTML (such as HTML 2.0), which is recognized widely, to code all essential content and navigation. If enhancements are used, care must be taken to check the pages in a no-frills browser to ensure that they are still readable.

Allowing for Minimal Scrolling

Related to the issue of user tolerance for Net lag is the amount of scrolling demanded by a page. Nielsen reports the dismal statistic that "only 10% of users scroll beyond the information that is visible on the screen when a page comes up" (Nielsen "Top Ten"). While this percentage shows a slight increase in recent studies, it still indicates that users will choose a link from the first screen's choices rather than take the time to see if other choices farther down the page might be more useful. For this reason, the general guideline is that a page that requires no more than three screenfuls of information is still a sound one. Also, every page must be designed so that the important information is presented at or near the top.

Enhancing Readability

Human-factors research indicates that readers take 25% longer to read a text on the computer screen than they do on the printed page, which offers higher resolution and sharpness (Gould and Grischkowsky; Nielsen "Be Succinct!")[4]. While improved technology may mitigate this problem over time, there are still a number of options available to Web page designers. Simply including more white or blank space on the page can also offer the reader visual relief, and this can be accomplished by the use of concise text, shorter paragraphs and line lengths, and the <BLOCKQUOTE> tag around the entire page. The effects of using or not using this tag are apparent in Figures 8.5 and 8.6 at right.

More importantly, readers' tendencies to scan online texts must also be accommodated. In a recent study, Nielsen and Morkes report that 79% of online readers scan a page, with only 16% reading word-for-word (Morkes and Nielsen "Concise"). Readers must, therefore, be guided into noting important details, a goal that can be easily met by the use of mechanical cues, such as highlighting key words and using headings, subheadings and bulleted lists; and the use of rhetorical cues, such as writing succinctly and arranging ideas deductively.

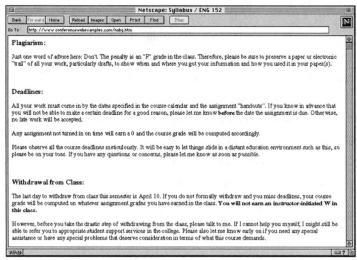

Figure 8.5 *Page Without the <BLOCKQUOTE> Tag*

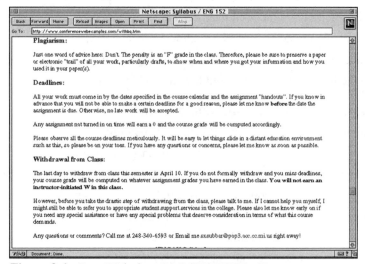

Figure 8.6 *Page With the <BLOCKQUOTE> Tag*

Considering Usability Issues
Relating to the Online Learner

While the issues described below are pertinent to all users, they are particularly important when pages are designed for online students who have to work on set tasks for an extended period of time within a unique framework created by a particular instructor.

Determining Purpose and Audience

All good design begins with good planning, and this is no less true for course page design. Online instructors must constantly examine two variables. First, what it is they are asking of their students on a particular page. Do they expect their students to complete a specific task in a specific sequence, or do they expect their students to discover for themselves meaningful relationships among various pieces of information? Second, which reading habits do they expect their students to bring to that particular task? Do they conceive of their students as browsers, users, or co-authors (to borrow Slatin's terminology, 153-69)? Different learning activities might benefit from different hypertext structures. For instance, if students were required to complete a task in a certain prescribed sequence, they would be using a different browsing strategy than if they were asked to read hypertext fiction like Joyce's "Afternoon."

Just because one is working with hypertext, one does not have to assume that *all* reading activity on the Web must be non-linear and all learning entirely self-directed. Navigational structures can allow users as much or as little freedom as the author intends. The design of every course page should reflect an awareness of both the instructor's purpose and the students' needs.

Building Predictability into Course Pages

Reading theorists and empirical studies of reading behaviors emphasize that "schemas," or mental representations of the structure of texts, are essential to comprehension (Charney 243-8). Readers come to texts with preconceived expectations regarding how certain material will be presented and where they will find certain kinds of information. If these expectations go unmet, readers become confused and they are less likely to remember what they read or to make meaningful connections among all that they read.

Some McPrinciples, so to speak, of course page design must be remembered. *All* the pages developed by the instructor for a particular class must have a similar look and feel to replicate the predictability of routines that a student typically encounters in the physical classroom. This sense of predictability can be reinforced in a number of ways. Just as it is important to compensate for the lack of contextual clues to reading hypertext, so too is it necessary to consciously provide rhetorical and structural clues to mitigate the disorientation the typical newcomer to hyperspace often feels.

A simple but very effective strategy is to use a consistent header and footer in *all* the pages developed for a particular course. The former offers the equivalent of a "You are here." notation on a map, while the latter can provide a way to exit the page gracefully (and gratefully!). Every page should also have a clear title to orient the reader who may access the page directly instead of through the home page. These strategies will allow students to focus on learning instead of on just trying to get their bearings.

Web users as a whole may be becoming less tolerant of one-of-a-kind design choices. Nielsen reports that "The Web is establishing expectations for narrative flow and user options" (Nielsen "Increasing Conservatism"). A consistent course page design can both save the instructor's time and facilitate student learning

Revealing the Overall Organization of the Course Pages

The cognitive importance of schemas also points to the usefulness of some kind of early graphical or hierarchical representation of the contents on any given page. Whether information is presented in a single long page or in a set of nodes, stu-dents-who read course pages primarily for information-will find it most useful to be given a sense of the overall organization. Research has shown that when read-ers are given a structured overview of the hypertext, they use the hypertext more efficiently and meet learning objectives better (Dee-Lucas 99). A site map, as illus-trated in Figure 8.7 (see page 120), can effectively explain how the entire course is organized, how the different components fit together. In a way, it can visually cre-ate a sense of community that might otherwise be lost in an online course.

Likewise, a table of contents is an excellent tool for orienting the reader to the contents of even a single page because it imposes a small measure of linearity upon what could be perceived as a confusing, discrete collection of information. Figure 8.8 (see page 120) shows how, even with individual assignments, students can be guided into reading all of the necessary information.

Designing for On-Screen Reading and Hard Copy

Just as instructors have little control over the order in which their students read hypertext course pages, so too do they have little control over how those pages will look on their students' screens or in hard copy. HTML, after all, is only a structuring language that allows the page designer to merely sug-gest a presentation. How the user will receive that presentation is beyond the designer's control. This is one good reason for designing pages that are sim-ple but useful.

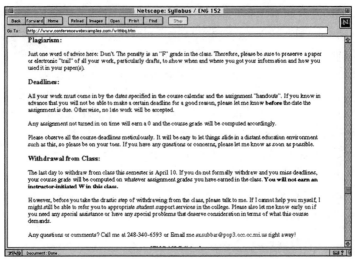

Figure 8.7 *Site Map for Pages Used in a Composition Course*

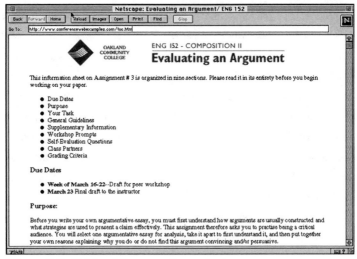

Figure 8.8 *Table of Contents for an Assignment*

Since most students tend to print out what they have to read—this is an observation supported informally by many online instructors—course pages must allow for comfortable reading both online and in hard copy. As discussed earlier, text and link colors and fonts must be selected carefully so as not to be lost in print. Pages must also be tested to see how different monitor sizes will affect the layout. On both PCs and Macs, resolution settings (found under the monitor "preference settings" menu) can be changed to simulate different screen sizes and the layout can be optimized accordingly. Figures 8.9 through 8.11 show how much of the same page appears on a single screen on three different monitors. Therefore, the instructor cannot assume that a "screenful" is a universal constant.

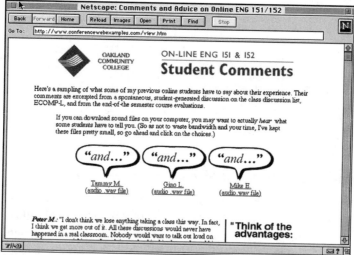

Figure 8.9 *(Macintosh) Screen Set to 640x480 Pixels*

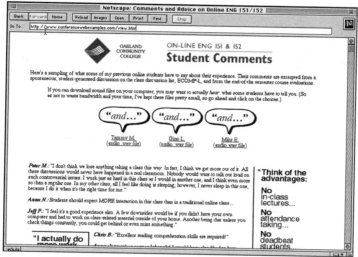

Figure 8.10 *(Macintosh) Screen Set to 800x600 Pixels*

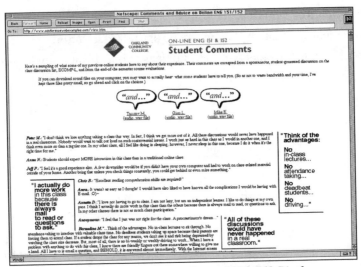

Figure 8.11 *(Macintosh) Screen Set to 1024x768 Pixels*

Determining an Optimal Size for a Course Page

While it is true that downloading a single file is less time-consuming than navigating a series of files, pages that require a lot of scrolling to read can become mind numbing. At some point, readers might become weary of all the seemingly endless information available and simply stop reading. Shorter pages might help ensure that more of the presented material is read. In addition, since it is hard to determine online page breaks, shorter pages might also allow students to print out just the segments they need. Most importantly, a short page might draw more attention to the information it contains and give it an aura of completeness that would not be perceived if this page were part of a long file. Instructors must, therefore, carefully determine the length of each page and the rationale for linking to separate nodes as opposed to including all the information within a single page.

Care should also be taken to provide enough contextual information within every page so that no matter which path the reader takes to a particular node, he/she gets the necessary content without having to refer back to a previously visited node. This, of course, leads to other design decisions regarding which texts and links must be repeated.

Constructing Useful Navigational Aids

The importance of predictability comes into play again in the construction of links to further information. The way a link is identified can hinder or aid learning. Simply asking a reader to "click here" generally serves no useful purpose and might only reinforce the novice's tendency to seek out what are often described as "low cognitive-load browsing strategies" whereby serendipity is all. Therefore, links must be phrased in a manner that allows the reader to predict what might be found when the link is selected. Too, if pages will be printed out, navigational links will be lost if not phrased properly. Figure 8.12 presents some examples of effective and ineffective phrasing of links.

The problem of link phrasing arises again if the user decides to print out pages. Figure 8.13, on page 124, shows how the URL is entirely lost in the printed version of the page when the designer does not include it in the sentence containing the link. The user will have to go back into the original page and click on the link, instead of being able to go directly to that node. Adding the URL, as in Figure 8.14, to important links (for example, <MAILTO> or any outside nodes) will make the addressing information accessible even when the users are not online. Link titling (the TITLE attribute within the <A HREF> tag) might also be used to aid navigation, especially when the link anchor or its surrounding context may not be sufficiently self-explanatory.

Effective or Ineffective?	Example of Sentence with a Link	Comments
Ineffective	"If you would like to learn more about web usability issues, <u>click here</u>."	Link does not indicate where it leads to or what information might be found there. The idea of "here" is itself relative on the web.
Ineffective	"<u>This link</u> contains information about web usability issues."	Web users scan the text. The phrasing of this link, however, requires that every word be read—something that the average reader might not be patient enough to do.
Effective	"Addressing <u>web usabiltiy issues</u> is critical to good design."	Readers can predict what is available at the node the link points to and why it is important.

Figure 8.12 *Effective and Ineffective Link Phrasing*

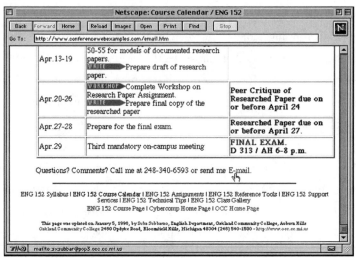

Figure 8.13 *Printed Version of Page Containing a Cryptic Link*

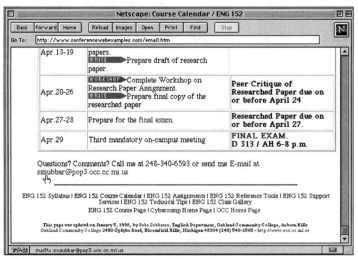

Figure 8.14 *Printed Version of Page Containing a Useful Link*

As with so many Web features, link titling is not supported universally yet, and the instructor must not assume that link titles will be received identically by all users. Also, the use of a link title will typically add about 0.1 second to download time. Still, as no distortion of the page layout will result, even if a browser does not support link titling, instructors should consider using this feature for at least what they consider to be the important links on a page.

Building Trust

Last but not least is the issue of trust. Whether an instructor teaches in the physical classroom or online, he/she is committed to creating an environment that will maximize learning. Clarity, consistency, accessibility, and fairness are some of the qualities that students expect-and online instructors can meet these expectations to a large extent by designing course pages that are truly user-oriented.

References

Charney, Davida. "The Effect of Hypertext on the Processes of Reading and Writing." In Cynthia Selfe and Susan Hilligloss (eds.). *Literacy and Computers: The Complications of Teaching and Learning with Technology.* New York, New York: MLA, 1994, 238-263.

Conklin, Jeff. "Hypertext: An Introduction and Survey." *IEEE Computer.* September 1987, 17-41.

Dee-Lucas, Diana. "Effects of Overview Structure on Study Strategies and Text Representation for Instructional Hypertext." *In* Jean-Francois Rouet, et al. (eds.) *Hypertext and Cognition.* Mahwah, New Jersey: Lawrence Erlbaum, 1996, 73-107.

Gould, John, and Nancy Grischkowsky. "Doing the Same Work with Hard Copy and

CRT Terminals." *Human Factors* 1984, 323-337.

"GVU's 8th User Survey." *GVU's WWW User Surveys.* 1997. <http:// www. gvu. gatech.edu /user_surveys/survey-1997-10> (10 June 1998).

Morkes, John, and Jakob Nielsen. "Concise, SCANNABLE, and Objective: How to Write for the Web." *Sun Microsystems,* 1997. <http://www.useit.com /papers/Webwriting/writing.html> (10 June 1998).

Nielsen, Jakob. "Be Succinct! (Writing for the Web). " *Jakob Nielsen's Alertbox for March 15, 1997.* <http://ww.useit.com/alertbox/9703b.html> (3 June 1998).

—. "How Users Read on the Web." *Jakob Nielsen's Alertbox for October 1, 1997.* <http://www. useit.com/alertbox/9710a.html> (3 June 1998).

—. "The Increasing Conservatism of Web Users." *Jakob Nielsen's Alertbox for March 22, 1998.* <http://www.useit.com/alertbox/980322.html> (3 June 1998).

—. "The Need for Speed." *Jakob Nielsen's Alertbox for March 1, 1997.* <http://www. Useit.com/ alertbox/9703a.html> (31 May 1998).

—. "Top Ten Mistakes in Web Design." *Jakob Nielsen's Alertbox for May 1996.* <http://www. useit.com/alertbox/9605.html> (2 June 1998).

Slatin, John. "Reading Hypertext: Order and Coherence in a New Medium." In George Landow and Paul Delaney, (eds.) *Hypermedia and Literacy Studies.* Cambridge, Massachusetts: MIT Press, 1991.

Endnotes

1. This phenomenon has been repeatedly observed in the author's own classes. It has also been debated in the electronic forum, Online College Classroom. For more information about

this forum or to access archives of relevant discussions, visit http://leahi.kcc.hawaii.edu/org/occ/about_occ-l.html.

2. The average factory setting for Macs and PCs is 1.8 and 2.5 gamma, respectively. Since gamma settings affect the brightness and contrast of a computer's display, their differences across platforms present a challenge to Web designers.

3. Other time wasters (at least from the user's perspective) are elaborate graphics to depict the site's being under construction, graphics that advertise the availability of a particular browser version, and graphics that advertise the software with which the page was created. Even if this information is relevant to the site, it can be presented verbally without burdening the user.

4. The average screen resolution is only 72 dpi, while the average page resolution ranges between 300-600 dpi.

CHAPTER 9

Teaching with Technology: Constructivism at Work

Mary Beal

Asst. Prof. Instructional Technology & Telecommunications

Western Illinois University
Macomb, IL

In the old days we used to talk about the teacher as expert, the "sage on the stage." There were a whole lot of passive students who were kind of like baby birds and the professor dropped the worm in their mouths. Now, however, professors in many fields are expected to serve as a guide-on-the-side, helping students learn how to analyze and synthesize information.

—Chere Gibson, University of Wisconsin

BACKGROUND

The Western Illinois University College of Education and Human Services (COEHS) received $9.2 million in 1998 to improve student distance education through its Center for the Application of Information Technology (CAIT). The COEHS hopes to reach the following primary objectives:

- To prepare students (there has been a 10% increase in the student population) for a future permeated with technology;

- To provide the university with innovative instructional materials that serve a variety of student learning styles;

- To teach students how to filter the burgeoning amount of information available to them;

- To promote collaborative learning and critical thinking.

As the COEHS funding situation began to improve, the administration created the Instructional Technology and Telecommunications (ITT) department, with its primary objective being to teach undergraduates and graduates, as well as pre-service and in-service teachers in the above areas. The ITT department, formed in 1996, was created in response to an increased need for technology training for students, teachers, and employees working in the field. Now more than ever, there is a need for a work force that is capable and knowledgeable about technology and the many ways it is being used today.

The ITT department, whose goals match those of the COEHS and the university in general, specifically aims at training in the areas of new and existing technologies to achieve educational goals in a wide variety of instructional settings, such as education, business and industry, government, health-related fields, and training. Students learn to design, develop, produce, and evaluate effective and efficient materials for instruction. The department utilizes internships and school-to-work opportunities for students in addition to working closely with business and industry in order to train students to meet the increasingly more rigorous demands of the workplace.

The ITT department currently has over 150 graduates and 50 undergraduates served by nine faculty. With an 80% entirely new curriculum, new course preparations are common—often faculty begin with only a paragraph or a sentence long course description. It is within this setting that I began teaching at Western two semesters ago and designed many new courses, one of which was project management for undergraduates. In it, I chose to use a constructivist approach to learning.

From my own experience working with the Oregon Department of Education for three years, I knew the constructivist approach means, among other things, modeling behavior for students and giving them the opportunity to work hands-on on their own projects. It means engaging students in the learning process and allowing them to discover and construct their own meaning out of their experience, using 3-D objects, models, computer-assisted instruction, and workplace settings to simulate real-world situations for their learning. In constructivist learning, students can be actively involved in the use of interactive e-mail, chat-rooms, bulletin boards, and Web browsers to further their learning.

TEACHING PROJECT MANAGEMENT

In teaching project management at Western Illinois University, I decided to employ what I knew about the constructivist approach to design and develop a class where students would learn to manage their own projects. My methods seem to have been effective. Student evaluations of their experiences were positive and, in addition, the local newspaper heralded our work and the accomplishments of the class. The dean of COEHS also was pleased. He wrote to the class and me a four-page letter commending their accomplishments.

As I examined the equipment that we had at our disposal for the students to use, I knew that partnerships of some sort would be necessary in order to provide the class the real-world atmosphere that I was seeking. In order for the students to have a full understanding of project management, two partnerships were arranged. In one, the students, in return for access to better equipment, would manage the university-wide television station. In the other, students would design a kiosk for the dean of the COEHS's office in return for funding for equipment. The pressure mounted for the students when a professional ENG camera and 1/2" editing equipment were delivered for their use.

Both the projects were highly visible. The administrators made it clear that the work was important to them, as did CAIT when the students took responsibility for the television station. The students were given the work of adults and I was proud to see them acting as adults would when faced with big responsibilities.

To begin, the students were asked to bring their resumes to class. I took this opportunity to critique their resumes and describe the portfolios of the work they would each be doing prior to graduation. From examining each other's resume, and from their own knowledge of their peers' reputations, they chose two team leaders. We called them the project managers for the two projects: the College of Education Station (referred to as COES-TV) and the kiosk project. Everyone in class was held responsible for some aspect of each team's workload. Each student wrote and signed a contract, which stated the work they agreed to finish by the end of the class to receive an "A." Class members chose roles of videographers, graphic artists, photographers, and production assistants as needed, and they had deliverables that they were responsible for by the end of the semester.

As the students decided on their functions, I developed a list-serve by which they could communicate with one another, if needed. They were allowed to e-mail me or call me anytime as questions arose during the class. Throughout the semester, when they researched the kiosk set-up and content, both the telephone and World Wide Web were used.

For the initial meetings with the class's two clients, I spoke to the students some about how they were to behave (professionally) and how they might dress. It was decided that the project managers would be the main spokespeople for the groups. The meeting with the dean and the department chairperson regarding the kiosk was very tense. The students were exceedingly nervous. The dean and the chair treated them with much respect. Work plans were made, interim deliverables were indicated, and the students were given a budget of $1500 to manage. From that initial meeting, students had enough information to apply what they knew, as I guided them in the utilization of Microsoft Project to develop GANTT and PERT charts in order to plan their time for each phase of their projects: analysis; design; development; implementation; and evaluation.

THE ANALYSIS PHASE

The class learned to analyze their audience and determine who, in fact, they were serving. A poll was taken around the university in order to find out who watched the television station. Students placed questionnaires in dorm room mailboxes. A pizza was offered as the prize for a drawing from all returned questionnaires.

At the same time, all departments within COEHS were contacted and invited to submit materials for broadcast and/or inclusion for the final kiosk. They were also given the option of having one of the teams produce materials for broadcast. Thankfully, many of the departments had materials already available on their Web site or on videotape that the teams were then able to incorporate into the campus broadcasts and also use as kiosk content.

THE DESIGN PHASE

The class met and decided upon a look for COES-TV that was then explicated in all graphic art and design to be broadcast. In addition, the kiosk design had to be decided and this had to be pleasing to the dean's office. Fortunately, a brand new logo had recently been distributed for the COEHS, and it was that logo that would be the hallmark for the kiosk as well.

THE DEVELOPMENT PHASE

Class members worked out logistics of managing the two projects and it was decided that everyone would learn everything possible. At first this seemed daunting to me, but the class adopted a train-the-trainer model. Two initial members learned how to program the digital TV channel, as well as learning analog video production. Then those two taught two more members, and so on.

I gave short lectures at the beginning of each class period. It was in one of those talks that everyone learned to make budgets using Excell spreadsheets. They learned graphic arts, digital still photography, how to create titles using Adobe Premiere, and how to build slide shows with Adobe Persuasion. They decided to use Persuasion to make a digital portfolio record of the class.

The project manager led the class in research on how to build an inexpensive, stand-alone kiosk. It was decided that Hyperstudio software and a Macintosh G-3 (with a mounted flat finger-touch mouse) for hardware would best fit the client's needs and a train-the-trainer model was used for everyone to become familiar with the system.

TEAMWORK

Luckily, our textbook, *The Little Black Book of Project Management* by Michael C. Thomsett (1990), covered many issues with which the students would become all too familiar. They learned many helpful skills, such as how to get along with others, how to work as a team, how to serve the client's ever-changing needs, and how to enlist help even though none may be offered.

IMPLEMENTATION

At the mid-point of the semester, the clients called a meeting with the students in order to see interim deliverables. The students once again were exceedingly nervous. They brought an architect's rendering of the kiosk framework, a draft promotional video about the project management class, a draft of an Adobe Persuasion presentation about the class, and their artist's renderings of what the "look" of the kiosk and COES-TV would be. The director of CAIT and the dean's office were very pleased with the interim progress. The work was peer reviewed at this time and formatively evaluated by the clients as well. Students were expected to make revisions in their work, which was subject to their client's taste and wishes.

The students finished their work by the end of the semester, on time. Throughout the semester, accountability was very important. Students could not miss class any more than they could miss a paid job. They were asked to call me or one of their peers if they would not be there. Only one student had trouble with attendance, but this same student showed remarkable progress and growth in his ability to be responsible throughout the semester, and I am hopeful for his future.

The skills the students have learned appear to be very transferable to the workplace. Already, several of the students from the project management course have received interest from potential employers. In addition, I believe that pure constructivism allows the professor to be a "guide-at-the-side" and utilize technology while putting the student's learning first. As S.W. Gilbert phrases it in his article, "The proper integration of technology results in changes not only in presentation and pedagogy, but in epistemology, in course structure, and content." I believe that teachers—in order to fully transform their classrooms to be more fully student-centered, lively, and appropriate to the information age of today—will need to translate their classrooms into meaningful experiences that replicate the real-world environment for their students.

Already, the local newspaper wrote an article about the accomplishments of this class. Also, the dean of COEHS wrote the class a four-page letter commending their work.

References

Chere Gibson, association professor, Continuing and Vocational Education, University of Wisconsin, Madison at http://madison.wlu.edu/~kobersteink/biblio.htm, http://www.wlu.edu/~joverhol/newchalk/.

Gilbert, S.W. If It Takes 40 or 50 Years, Can We Still Call It a Revolution? *Educational Record*, summer 1994, 75(3), 19-28.

Thomsett, Michael C. *The Little Black Book of Project Management*. New York, NY: Amacom, 1990.

CHAPTER 10

Positioning Web-Based Learning in the Higher Education Portfolio: Too Much, Too Soon?

Steve Flowers

Steve Reeve
MBA Program Managers

University of Brighton
East Sussex, England

INTRODUCTION

The main features of much that has been written about the role of information technology (IT) in higher education are its strong future-orientation and its often overwhelming optimism. There is, of course, nothing wrong with such optimism, but perhaps there should be more rigorous investigation of the basis for such a positive attitude. There may be a danger that much of the immediately foreseeable advantage might eventually be replaced by a much starker long-term outcome of what may be a seismic paradigm shift. If some scope for a more jaundiced analysis is allowable, the future of virtualized higher education might more closely resemble the following scenario.

Picture a call center; there are clusters of staff in front of screens. In pleasant tones, these people are working through the programmed information being presented in front of them. They are titled courseware managers, and are essentially a mixture of graduate assistants earning part-time wages and professional call center personnel. At the other end of the line are students, seeking advice on assignments, asking for further explanation, navigating through courseware structures. In a different section of the room, others are busy grading work and giving evaluation and assessment feedback online.

The courseware itself, and the expert systems being interrogated by the staff, have been leased from one of the major prestigious providers. This university has many thousands of students from all over the world, and is licensed to confer the degrees of several leading academic institutions. It is neither regional, nor prestigious in research terms; it is, in fact, one of the standard "distributor" universities (the fate of many higher education institutions worldwide). It is certainly relatively cheap to operate, and there is no doubting its efficiency both in terms of return on capital and academic added value. Its academic (in the old sense of the word) staff were long ago scattered to the four winds; some now earn their living in the major courseware development houses, others are researchers at category 1 global innovation centers, a few stayed on as call center managers. Most, however, joined the vast army of self-employed knowledge consultants, picking up private contracts as best they can in a fiercely competitive environment.

Of course they regret the passing of the old ways and often refer fondly to the teaching times, when people used to get together in a single place to communicate and share information, emotion, and learning. Such blatant antisocial resource wastage was obviously ripe for clear-out, but they can't help looking back with fond memories nonetheless. When the Great Rationalization came, it took no hostages. "Get online or get out!" roared the resource managers as great swathes of physical contact teaching were revealed as hopelessly uneconomic in the face of virtual competition. "Upgrade or ship out!" exhorted the banners hung clumsily around the rooms of recalcitrant lecturers. "Virtualize or die!" screamed the innermost e-mails of rapidly collapsing academic universes.

And so it was. As the universities and colleges locked together in mortal combat, the technological escalation ran riot; online learning was the only ammunition, and those who operated online faster, more widely, more comprehensively, and more innovatively gained the advantage and accreted the market around them. With only a finite number of financially capable global learners, there clearly had to be losers. Easy to say in hindsight, but higher education came to regret the no-holds-barred, compete-on-technology leitmotif that had brought it to the brink of systemic collapse.

Still, that's all in the past. The Coordinated Governments and Large Companies Initiative brought a sense of order, reorganized the field, and generated the sense of peace and cohesion now evident in the virtual edu-world.

This chapter will attempt to deal with certain major themes that are propelling the learning technology revolution. It seeks to clarify and refine a wider-scale view of the prospects for higher education, and individual institutions within higher education, and will present a series of key issues that need to be addressed by these institutions. In developing the critique, issues will be examined from a variety of contrasting moral, social, and technological viewpoints, with the intention being to move beyond the more usual and immediate discussions around the operation and effectiveness of delivery systems.

COMPETITIVE ADVANTAGE FROM LEARNING TECHNOLOGIES?

Borrowing from the lifecycle type of analysis found within the marketing domain, higher education could be facing a major shift of ethos, structure, and process as it moves into a new stage of its evolution. Furthermore, as institutions begin to compete more aggressively with each other, it is likely that they will attempt to use technological delivery or online learning as a source of competitive advantage within their self-perceived market. However, the very act of attempting to create and exploit such an advantage could lock the whole system into a new paradigm or infrastructure that could have unpredictable long-term implications. While there are very few organizations that have achieved a sustainable, first-strike competitive advantage (e.g., American Airlines and its SABRE system), the replicability of IT solutions has meant that it has proved to be notoriously difficult to sustain such an advantage.

It must be acknowledged that the enthusiasm for virtual learning is certainly very strong within certain sections of higher education and the discussion of the implications of IT for education as a whole (Daniel 1998; Twigg and Oblinger 1996; Hague 1996; Kay 1995; Flowers 1993; Leggat 1993; Searl 1993) has been underway for some years. However, the often technical language, general tenor, and highly positivist approach adopted by much of this literature not only serves to make it largely inaccessible for many non-IT-literate academics, but also means that it does not speak of today's problems but rather attempts to provide tomorrow's solutions.

It is perhaps significant that a more sober, complex, picture of the issues is presented by those actually engaged in the development of learning technology (Swanton 1998; Maki and Maki, 1997; Vetter and Severance 1997; Turoff 1995). The strong positivist tone previously identified is tempered in this literature by a more reflective, more tentative approach that recognizes the significant challenges involved in the development of Web-based education. Notwithstanding the enthusiasm for the development and adoption of learning technologies, challenges to this position have begun to emerge, celebrating, exposing and attacking the strengths of the old technologies of the textbook and lecture (McKenzie 1997) and weaknesses of the new Web-based approaches (Hecht 1997; Aujla, 1997).

However, it must be recognized that the widespread promotion and adoption of online learning within higher education is being driven forward by a powerful combination of forces that are operating, in addition to the more acceptable moral rationales. Technology advocacy in itself is clearly important, but so too are distinctly political motivations. It is likely that funding from private corporations, jockeying for position within the academic community, potential economies, opportunities for organizational reengineering, fear of apocalyptic rhetoric, political standing, and funding inherently attached to innovation are all factors that combine to provide momentum for the widespread adoption of learning technologies.

It may become incumbent on institutions to examine their role much more closely and define where they see themselves in a future where large-scale technological fixed cost and electronically mediated delivery are the norm. Within this environment, it is certainly timely to examine in great detail the potential advantages or problems that may manifest in the educational philosophy arena. If the contextual environment dictates large-scale transference to online systems as the predominant model, what effects might there be for education itself?

ADOPTION RATIONALES FOR LEARNING TECHNOLOGIES

In general terms, there appears to be three leading rationales for the pursuit of online approaches: One exists firmly in the moral sphere and concerns issues of access; the second is to be found within education theory and concerns effectiveness of delivery; and the third concerns the role of IT as a mechanism by which learning and teaching can be enhanced or automated within higher education.

Access

The access question sits within a wider social imperative. The overarching assumption here must be that the greater the participation of a given population in higher education, the more productive, innovative, and civilized that population becomes. In what sense, therefore, does a new or adapted online course increase access to higher education? What kind of access widening might be occurring?

For example, the main thrust could be to include students previously disenfranchised by physical, social, and economic constraints. However, it could mainly concern the classical distance-learning arena where students are geographically distanced from mainstream provision. A rather more sinister tone is also detectable as institutions compete, that of the "capture" of other institutions' students in a beggar-my-neighbor approach. Another less controversial reason could be to free up staff within a standard program to engage with the greater difficulties experienced by an increasingly diverse "non-standard" student body.

Clearly, before commencing design, purchase, or implementation of educational technology systems, such specific questions should be addressed. These specifically targeted questions are not often raised, and the importance and value of debate among all the academic stakeholders can be lost. The absence of such debate can become problematic two or three years later, by which time second or third wave lecturers are directly involved in the maintenance and success (or otherwise) of suites of online program. This is not to suggest that initial pioneering online education is not needed. Obviously their catalytic action is imperative if institutions are to move forward, but the overall debate should be joined at the outset rather than when it might be too late.

Furthermore, the access question itself can be problematic. If far greater access to education is provided (as is the case in standard open or distance program) but large numbers of students drop out, experiencing a sense of failure and disenchantment with education, how far should that be measured as a success? At issue here is whether there might be a significant difference between the attrition rate associated with classic distance/open techniques, and that associated with Web-based learning. Furthermore, in qualitative terms, what is the potential for the Web-mediated experience to be more satisfactory and meaningful than the traditional open learning experience and, thus, escape the "wastage" that so often defines that system?

There is a danger that an immediate answer to this kind of question is both affirmative and based on assumptions rather than any clear evidence. Despite the multi-level linking, search engines and vast amount of data within the Web universe, would an individual student be any less "alone" than a classic open learner? Navigating as an individual through the vastness of the Web, with all its bewildering potential, may produce a far less satisfactory learning experience (both in cognitive and affective terms) than may be obtained within the traditional, bounded, closed curriculum world of conventional face-to-face education. One issue is clearly linked to the other; if the conversion to online systems is made in the name of greater access, we have to be reasonably sure that those gaining access stay with it, and that those who might previously have received a face-to-face educational context do not suffer an overall reduction in the quality of the experience.

Hopes for the engagement and success of previously uninvolved students might be confounded by the self-selection demonstrated by the traditionally successful student, in the face of all the expectations. Web-mediated education has the potential to provide for all those in the past served badly by a conventional curriculum (i.e., quiet, shy, minority, multicultural students, etc.) but evidence of performance is still unclear. Dziuban (1998), for example, suggests that interim data points to the traditionally successful students (outgoing, strategic, field dependent) as the self-selectors who are currently succeeding within online education. Such research might point the way forward toward a more niched and customized learning program, essential to genuinely provide access but increasingly costly in terms of design and operation.

EFFECTIVENESS OF DELIVERY

The second major theme, concerning the effectiveness of delivery, can be divided into two forms: first, an economic argument—more students can learn at a lower societal cost in terms of resources consumed per student; and second, that students can learn more effectively using the new technology available. Taking the second of these, much empirical and evaluative work still needs to be done before overtly propagandist justifications for expenditure and implementation should be heeded. It is also important to compare the theoretical and empirical work that does exist against the claims being made for the new technological approach. One particularly large theoretical challenge concerns the richness and value of the learning.

The evolution of technology-based education has clearly followed the format for distance learning, and essentially shares many of its principles. One finds an emphasis on individualization, serial approaches, surface methodologies, "knowledge as parcels," rational (i.e., not emotional) learning, autonomy, and the ability to "stand alone." This is not to suggest that the delivery mechanism for open learning has not undergone a change—as Net-based, more or less interactive, and closer asynchronous contact structures have evolved as online models—but rather to serve as a reminder of some of the embedded principles within the "DNA" of current state-of-the-art approaches.

Such infrastructural constraints and direct relationship to more elemental forms of distance learning have ensured that computer-based learning has been specifically both disseminationist in tone and surface/serial in character. This history still dictates the shape of much specifically written educational material, which contains the worrying potential to trap students on the lower levels of the taxonomic ladders found within educational philosophy. A kind of utilitarian calculus also exists, whereby a "second best" argument (in the theoretical economic sense) is proffered, in that shallower and less effective individual learning experiences might be traded off against increased access overall, or that the experience impoverishment might be outweighed by the cost savings.

Of course, great strides have been made in terms of increasingly sophisticated courseware to combat such deficiencies, but the further the attempt to enrich courseware goes, the more expensive the whole enterprise becomes and the more serendipitous becomes the students ability to access such riches (depending on standards of equipment owned, leased, or borrowed). Educators then find themselves embroiled in policy discussions as to whether their systems should be cheap, accessible, robust but basic, or complex, rich, leading-edge, and fragile. Before long, moral points about the nature, standard of income, and expectations of potential students become key to the discussion. Institutions have to weigh involvement with private corporations, technological "lock in," and capital for leasing schemes against independence, lower functionality, and operational cost. These are clearly non-core issues for educators that will inevitably lead to a dilution of the efforts that should more appropriately be applied to the core activities of learning and teaching.

SOCIAL LEARNING?

This is a particularly interesting area of research and debate, given that on one hand there is a clear disseminationist sense of controlled or programmed learning in the direct antecedents of online learning; while on the other, much is made of the postmodernist, anarcho-chaotic nature of "The Web," and the individual's relationship to it in terms of learning. Further compounding such debate is the elevation of Vigotsky as iconic philosopher to a Web medium, where all learning, by default, becomes a potentially social experience. Given such lack of clarity, it is important to understand and take cognizance of the nature of social learning within current higher education environments. It is also important to consider carefully some of the limiting features of Web-based approaches, in order to explore the creation of effectively educational social interactions within a Web-based educational context. It is perhaps ironic that the implications of social learning from more traditional educational domains are becoming associated with Web-mediated learning processes, lending an air of authoritative social learning to an educational infrastructure that may yet prove anything but social in actual operation.

Such socialization rhetoric has no greater impact than in the decision to transfer conventional courses to online format. When such a direct transfer of curriculum model is undertaken, one model predicated on personal contact, "known" parameters of behavior (Miller and Parlett 1974), informal social learning, and deeper/more holistic architecture is replaced by another. This replacement is predicated on individualized "virtual" contact (even with e-mail interactions, chat rooms, etc.), publicly knowable informal learning (given the necessary public domain of e-mail or Web space), still somewhat unclear conventions, and predominantly surface/serial courseware structures.

Whether the generation of the postmodernist notion of "multiple affordance" learning (Ryder and Wilson 1997) can really replace the genuine, if culturally specific, social learning experienced by conventional students remains to be questioned. Whether, and to what extent, traditionally field-dependent learners may genuinely engage with the multiple affordances on offer, is again a moot point.

Researchers and teachers working actively with online learning models are capable of producing statistical and anecdotal detail to demonstrate that social learning (or indeed socialization) is there. However, much of this is to be treated with some caution, as particularly "hit rate" data is not comparing like for like. It is still very hard to calculate how the statistically monitorable chat room, teacher/student, and student/student interactions can be compared meaningfully to the degree of "hidden curriculum" activity undertaken by students on conventional courses. Equally difficult to interpret are figures that compare online interaction with previous bulletin-board type interactions within conventional courses. Certainly, more carefully monitored micro-scale evaluations may even refute the interaction "logic" (Smith 1998) by demonstrating fewer rather than more numerous contacts. A further difficulty that compounds the problems of comparison between traditional and online interaction is the use of assessment to motivate students to make contributions to a discussion group, thereby inflating the level of interaction and distorting the role such discussion groups may have within a learning environment.

Within the United Kingdom, the groundbreaking work on learning orientations and strategies pioneered since the mid-1970s (Marton and Säljö 1976; Pask 1976; Entwistle 1988; Biggs 1985) has become an authoritative view among educators, educational psychologists, and field researchers. If this current consensus on learning approaches is taken on board, it implies a relative inferiority of serial/surface learning compared to deep/strategic orientations. Studies conducted specifically concerning open/distance learning by Harper and Kember (1989), within higher education by Richardson (1994) and Scouller and Prosser (1994) continue to validate the original work, confirming the effectiveness of deeper learning orientations.

It also clear that the work driving the current interest in collaborative learning, based on Lave's situational approach and particularly Vigotsky's writings on the pre-eminence of the social within learning, implies engagement with the deeper/holistic orientations mentioned above. Where the two might part company concerns the degree to which students might be aware of, and capable of, manipulating, their own learning orientation. There seems to be an assumption that, of itself, the presence of the Web somehow enhances a student's ability to construct their own meaning and learning. "The designer has lost control, but the medium has gained credibility. We no longer have to contrive interactive 'lessons' and exercises. The real world is waiting on the other side of the terminal" (Ryder and Wilson 1997).

Learning approach theory, however, would suggest that results flowing from such engagement may well simply demonstrate the Paskian notion of "globe trotting" at a surface level. How important could the Web be in aiding students to consciously structure their own deeper, strategic learning? The answer must lie in the possibility of harnessing the potential of the Web to a specific program designed to help a student learn how to learn successfully.

Thus, not only will students have to approach their learning and navigation differently, so too will tutors. One of the less understood considerations here concerns the professionalism and "subconsciously" acquired and utilized skills of the good educator to compress the variability of a class-based cohort. Once online and in a virtual world, an unmanaged stretching out of variability could take place, with the autonomous, strong students flourishing while dependent, weaker students flounder. Systems will have to be organized to deal with such unleashed variation and complexity, previously coped with by skilled teachers as part of the job, which in itself suggests a whole new mode of educational working. Already, the anecdotal tales of being "chained to email" are echoing around from the institutions furthest down the online road.

Even if institutional authorities can make the psychological shift to the lecturer as someone who designs courseware, deals with students and assessment, and mediates all such activity from a work station, a massive operational change is implied. Where institutions are forced online by competition, or politics, and a previous infrastructural model is impacted by the above, the medium-term damage to staff and students cannot be underestimated.

One possible casualty of this change is the vital part that academics play, at their best, in imbuing a sense of excitement into the academic process and providing the intellectual stimulation that is so important to students throughout their studies. Another important role that is often overlooked is the part teachers play in providing interactive frameworks of the evolving landscape of a discipline together with an introduction to language of discourse employed within that discipline. While taken for granted within the process of higher education, these are key features of the traditional model that are very difficult to pass along in an online form.

The big question, therefore, must concern whether a virtual, online program (with concomitant links into the Web) has the capability to teach students how to consciously develop deeper, more strategic learning approaches. If this is possible, it requires a substantial input of traditional contact. If it is not be possible, students in such a program may never progress beyond surface/serial approaches and are left to float, clinging to their courseware, isolated, and confused in the vast ocean of the Web. Another possibility is that the courseware and program management of a virtual process are in themselves capable of generating such meta-level awareness among students. Such a possibility currently seems remote.

If the paradigm shift occurs, and macroeconomic circumstances force a universal switch to virtual higher education environments (as opposed to educational philosophy driving such change), the implications would appear to be as follows:

- Individuals (not class members) will be engaging in the higher educational project

- Personally constructed learning will be the main learning mechanism

- Shifting virtual communities will provide the social context

- Non-dynamic intellectual frameworks will provide the over-arching structures

- The written word will become the primary means of communication and discourse

- Web-navigated learning will form the primary source of engagement

Clearly the above presents a radically different view of both learning and curriculum from the current ancient traditions prevailing in most institutions, particularly for younger, full-time students. If such students are not prepared, enabled, and capable of dealing with this radically different form of education, they will founder. Furthermore, to the extent that staff are not aware, conversant with, and culturally attuned to such new learning frameworks, the danger to future generations of learners becomes even more acute.

The very role of higher education within society is also at issue here, as clearly there is a role beyond that of merely educating individuals. The recent government report (Dearing 1998), produced in the United Kingdom, directly tackles the nature of the societal learning with which students should engage the betterment of civil society and democracy. This necessitates an emphasis on group behavior, peer engagement, debate, emotion, and maturity as citizens. Such educational goals may become ever more unobtainable as the educational infrastructure fragments and atomizes into an individually predicated virtual world.

THE ECONOMICS OF LEARNING TECHNOLOGIES

No one seems clear about the new economics of significantly virtualized higher education. There is clearly an attraction to the talk of economies of scale, lower resource cost per student, increased flexibility of course provision, and change. However, such discussion resembles the kind of financial discussions, which have surrounded distance and open learning. It remains difficult to find individual college or university experience on the ground where open learning has resulted in any cheaper than traditional education. It has become almost axiomatic in open learning circles that either open learning costs the same as standard tuition (it simply permits wider access) or it becomes "cheap" only at the expense of the infrastructure (curriculum development, learning support, fixed cost resource, etc.) usually deemed necessary for high-quality education. Combined with expected attrition rates, the absence of such architecture may allow a marginal costing model to predominate. In these terms, however, such space is more often inhabited by commercial or quasi-commercial activity than by true "academic" distance study.

Furthermore, in the technological arena, much of the future cost is unknown. Those experimenting with online systems1 report increasingly frequent and increasingly expensive environment upgrades (with a great number of unanticipated administrative costs), problems with delivery capability and student reception discrepancies, costly downtimes, and a cost-versus-capability problem in terms of "diminishing returns" to learners. It is also clear that development activity and start-up cost support often comes from IT companies; IBM, for example, is providing equipment to colleges and universities as the basis for their online program. The involvement of such companies is now deemed essential in many of the larger scale, pan-continental virtual structures, e.g., Monterey Institute of Technology pan-Latin American cooperation and the European Union "EUROPACE" project. One-off, pump priming investments coming from the "outside" (lottery money, development funds, etc.) further cloud the true cost issue.

It becomes increasingly difficult to evaluate or compare in order to generate value for money calculations. Some institutions are opting for the lowest basic standards so as to ensure control over cost escalation, maintain continuous service (no downtime), and provide maximum access, but there is clearly a downside here in terms of the richness of the student experience. What price for "online correspondence courses?" There is a great temptation for institutions to move to a "distance learning" version of something that is currently popular, place it on the Net, and hope the money is available. This may well be one of the most harmful variants of the new online universe, where educational goals, philosophy, staff training, and student expectations are potential sacrifices on the alter of revenue.

The very definition of distance learning can become key in the moral debate about potential outcomes. Will the existence of online learning facilities from major higher education institutions necessitate the collapse of smaller, geographically dispersed colleges currently serving regional needs? What is now meant by regional? The same discussion seems to be held whether the facts at issue concern widely spaced, geographically defined distance education—for example, in parts of Canada, Australia—or what counts as over one-hour "commute distance" by car in the United States. The line between morally acceptable contingent casualty and unfair "piracy" is becoming difficult to interpret. The economics and politics of regional provision will no doubt form an important ingredient in future research into the effects of the new technologies.

While the discussion so far has been concerned with the economics of online learning from the provider, or university, perspective, it must be recognized that the move to such systems also imposes significant additional costs upon the student. The costs of accessing education in this form are significant and, unlike the traditional methods of access, are ongoing and will involve, at a minimum, access to a PC, modem, printer, a range of software, and an on-ramp to the Internet. While many students may own or have access to such technology and be able to pay the ongoing costs associated with its use, many may not.

Any system of higher education that is predicated upon the ability to use such significant assets will inevitably disenfranchise a large number of potential students from the least well-off groups within society. It is likely that the increased access claimed by champions for online education may be off-set by the exclusion of those groups who have most to gain from higher education but are unable to afford the admission fee, in terms of hardware and software, to the virtual university.

IT AS A MECHANISM FOR ENHANCING OR AUTOMATING LEARNING AND TEACHING

The third issue, the role of IT as a mechanism by which learning and teaching can be enhanced or automated within higher education, has within it a series of interconnected issues relating to the complexities of applying technology effectively within organizations. Within the debate over the use of IT to create the virtual university, no real position has yet been established over what defines the "appropriate" role of IT within the educational process. The initiatives underway range from the straightforward provision of textual materials that support traditional lectures, to more radical virtual-learning environments that make extensive use of custom-designed courseware and discussion groups, with other systems employing elements of both approaches. Many of the systems are experimental in nature and, as has been mentioned above, evaluation has been patchy. Such evaluation that has taken place has been inconsistently applied, with the result of benchmarking being impossible. It could be argued that many institutions are in the very early initiation/adoption stages with this cluster of technologies and that many of the key issues relating to its successful application have yet to be discovered.

The tone and content of much of the positivist rhetoric surrounding the potential of learning technologies appears to be following the profile associated with the application of technology to a previously intractable area. Within this technology positioning profile, the earliest reports will announce the technological advance but typically focus far more on the potential of the technology in the very near future rather than the meager achievements made thus far. These reports will typically underplay or ignore the complexities of using the technologies and will focus on their potential for facilitating a transformational change or paradigm shift. The next stage will be largely uncritical reports of its use within one or two pilot sites and will provide glowing accounts of the success of the implementation and the gains made with, once again, little or no discussion of the real complexities encountered or its limitations.

Stage three generally comes after the technology has been adopted by a much larger group of organizations. In this stage reports are usually far more balanced, citing both the benefits and limitations of the technology. It is no longer presented, or viewed, as anything more than another element within a portfolio of technological approaches, with its potential for facilitating a change in paradigm being constrained by a wide range of limiting factors. Stage four of the technology positioning profile is the replacement of the technology by a competitor. Stage five is its obsolescence and eventual abandonment.

It could be argued that this whole process of technology positioning is managed with the intention to move the market as quickly as possible to stage three in order to maximize returns from a technology before it is either supplanted or its weaknesses recognized. It is probable that higher education as a whole, in the United Kingdom at least, is still between stages one and two and we have yet to move to anything like wide scale adoption on learning technologies, although the pressure to do so is building fast.

RISK AND THE "CHASM"

The risks of this position are well known and the potential for sub-optimal outcomes in areas of great technological promise but little direct experience have been documented elsewhere (GSA 1988; Flowers 1996), yet the move towards the wholesale adoption of learning technologies appears unaffected. One insight into this adoption pattern may come from the work on the adoption of innovations undertaken by Rogers (1995) and others. While it is not proposed to discuss this literature in any depth, for the purposes of this chapter it is important to recognize that it proposes two things. First, the existence of a well-defined pattern in the way in which an innovation is likely to be adopted. Second, that adoption of the innovation will proceed through a predictable sequence of well-defined adopter groups, the first of which may be termed "early adopters" (including individuals who are willing to try out any new idea, often for its own sake, as well as those who will look for ways to make dramatic improvements by applying it to core activities). The others may be termed as "mainstream users" who are a more conservative group consisting of those interested in more gradual incremental change and the concrete benefits that can be obtained, as well as more skeptical individuals who are likely to be late adopters.

Applying this model to the adoption of IT in teaching and learning within higher education, it could be argued that much of the activity, and thus contributions to the debate, is flowing from the early adopter group. While this is not unexpected, for the innovation to succeed, those in the mainstream must pursue it, a move that is by no means automatically assured. Indeed, the difference in needs and wants are so great between these two groups that the existence of a "chasm" between early adopters and the mainstream (Moore 1991) has been proposed. It is into this chasm that many high-technology innovations have fallen in the move from early adopters to the mainstream. In the world of educational technology, a host of innovations, including interactive video, computer assisted learning, and computer based training have, arguably, all fallen into Moore's chasm.

Recent experience in this area (Flowers, Newton, and Paine 1998) has only served to emphasize the importance of managing the cultural and organizational changes, in addition to the technical changes during the successful implementation of learning technologies. All organizations have examples of technically excellent IT systems that were never used to their full potential as a result of poor management of their design or implementation. Indeed, it could justifiably be argued that the technical challenges facing universities as they move to adopt learning technologies are but a small adjunct to the much wider series of organizational and micro-cultural issues. These issues will ultimately determine the way in which the cluster of approaches that currently constitute learning technologies are used within higher education.

ORGANIZATIONAL DNA AND LEARNING TECHNOLOGY

The institutional variability within higher education is significant, with organizations varying across a wide range of factors, including their size, teaching or research orientation, degree of internationalization, number and type of disciplines, organizational culture, traditions, and geographical location. While the sum total of these factors—the DNA that defines higher education—is shared by all organizations, each institution's DNA profile will vary in many ways from every other profile. Universities are thus distinct individuals with unique DNA profiles, albeit belonging to the same family, probably having their origins in the same gene pool and sharing many similarities, both significant and superficial.

The implications of this analysis are that while some learning technology therapies will be effective across all organizations within the sector, many will not. Specific approaches will have to be developed that may, in turn, provide the basis for further generic therapies. While some organizations will engage in radical, high-risk, genetic engineering approaches, it is likely that many will prefer to apply tried and tested learning technology therapies that have known, and sometimes controllable, side effects. Over time, as with the project to map the human DNA, the immense complexity of the undertaking will be widely recognized, the transformational rhetoric will be toned down, and organizations within higher education will recognize that learning technologies are no quick genetic fix. Learning technologies should form part of a coherent, planned, long-term genetic therapy.

Given the genetic diversity within higher education, institutions should be clear about their own strategic and pedagogical goals for adopting technology-based learning. Debate should take place within institutions considering embarking on such a move, which should include all stakeholders, and local solutions adopted, if necessary. Comparative discussions should take place about which models serve which needs best, and which costs and benefits are realistically assessable, with the outcomes from such discussions being clearly enunciated and specified in terms of access, student growth/retention, effectiveness of learning, and financial expectations.

It should be expected that the adoption of such technologies will vary dramatically between discipline areas. Organizations should avoid a "one-size-fits-all" approach. Promoters and users of technological approaches should recognize that they may be involved in, to a lesser or greater degree, a risky pioneering activity in which many of the key issues regarding effective application await clarification and resolution, and that a "learning by doing" phase is likely to be required. The challenges of scaling-up successful pilot projects should not be ignored, and organizations should expect projects to be evaluated rigorously and side effects reported. Evaluation should enable progress to be measured as learning technology programs develop. Institutions should expect to be able to develop indices that allow traditional and technology-based approaches to be compared in a meaningful manner. Finally, institutions should not be afraid to abandon those learning technology therapies that are either unable to deliver on their promised benefits or have too many side effects and adopt either traditional or hybrid approaches.

CONCLUSIONS

This chapter has attempted to identify and examine some of the major drivers that are propelling the learning technology revolution in higher education. It has also sought to clarify and refine a wider view of the prospects for the adoption of learning technology within higher education. It has presented a series of key issues that must be addressed by institutions prior to making the leap to the virtual university. In developing the critique, issues have been examined from a variety of contrasting moral, social, and technological viewpoints, the intention being to move the debate beyond the more usual domain relating to the operation and effectiveness of delivery systems.

It is proposed that the majority of higher education is still in the very early stages in the adoption of learning technologies. Given this information, there is still time to initiate an inclusive debate within and between institutions and develop evaluation, adoption, and implementation strategies that recognize and build on the important "genetic" variation within higher education.

References

Aujla, A. (1997) Virtual university, virtually useless, *The Peak*, Vol. 96, Issue 6, June 9. Available at http://www.peak.sfu.ca/the-peak/97-2/issue6/editor.html.

Biggs, J.B. (1985) Metalearning and Study Processes. *British Journal of Educational Psychology* Vol.55, part 3, 185-212.

Daniel, J. (1998) "Virtually all you'll need to know," Guardian Higher Education Supplement, *The Guardian*, April 7, pii-iii.

Dearing, Sir R. (1998) *Higher Education for the 21st Century: The Dearing Inquiry.* http://www. open.gov.uk/dfee/highed/dearing.htm.

Draper S.W., Brown M.I., Henderson F.P., McAteer E. (1995) *Integrative evaluation: An emerging role for classroom studies of CAL.* WWW: http://www.psy.gla.ac.uk/~steve.

Dziuban, C., Dziuban, J. (1998) Reactive Behaviour Patterns in the Classroom. *The Journal of Staff, Program, and Organizational Development* (in press).

Entwistle, N. and Waterston, S. (1988) Approaches to studying and levels of processing in university students. *British Journal of Educational Psychology* Vol. 58, part 3, 258-265.

Flowers, S. (1993) Want It? Gopher It, Computer Guardian supplement, *The Guardian*, 18 August, p. 12.

Flowers, S. (1996) Software Failure: Amazing Stories and Cautionary Tales, J. Wiley & Sons, Chichester.

Flowers, S., Newton, B., Paine, C. (1998) *Creating a Faculty Intranet: A Case Study in Change, Education and Training.* MCB University Press, Volume 40, No 8.

General Services Administration (1988) *An Evaluation of the Grand Design Approach to Developing Computer Based Application Systems*, U.S. General Services Administration: Washington D.C.

Hague, D. (1996) From Spires to Wires, Guardian Education supplement, *The Guardian*, February 13, p. 2.

Harper, G. and Kember, D. (1989). Interpretation of Factor Analyses from the Approaches to Studying Inventory. *British Journal of Educational Psychology* vol. 59, part 1, 66-74.

Hecht, B. (1997) Net Loss, *The New Republic*, February 17, Obtained from http://magazines.enews.com/magazines.tnr/textonly/021797/.

Kay, A. (1995), Computers, Networks and Education, *Scientific American*, September, Special Issue Volume 6, No 1, p. 148-155.

Lave, J. Rogoff. B. (eds.) (1984) *Everyday Cognition.* Harvard: London.

Leggat, R. (1993), Student days on modem campus, Computer Guardian supplement, *The Guardian*, 4 November, p. 17.

Maki, W. S. and Maki R. H. (1997) Learning Without Lectures: A Case Study, *IEEE Computer*, May, Volume 30, No. 5, p. 107-108.

Marton, F. and Säljö, R., (1976) Symposium : Learning processes and strategies -II. On qualitative differences in learning. Outcome as a function of the learner's conception of the Task. *British Journal of Educational Psychology* vol. 46, part 2, 115-127.

McKenzie, J. (1997) In Defense of textbooks, Lectures and Other Aging Technologies, From Now On,: *The Educational Technology Journal*, Vol 6, No.8, May. Available at http://www.fromnowon.org/may97/defense.html.

Miller, C.M.L., Parlett, M. (1974) *Up to the Mark: A Study of the Examination Game*. Society for Research into Higher Education: London.

Moore, G. A. (1991) *Crossing the Chasm*, Harper Business: New York.

Pask, G. (1976) Styles and Strategies of Learning. *British Journal of Educational Psychology* vol. 46, part 2, 128-148.

Richardson, J.T.E. (1994) Mature students in Higher Education: I. A literature survey on approaches to studying. *Studies in Higher Education*, Vol. 19, no. 3.

Rogers, E. M. (1995) *Diffusion of Innovations*, (4th edition), The Free Press: New York.

Ryder and Wilson (1997) *Affordances and Constraints of the Internet for Learning and Instruction*. Available at http://www.cudenver.edu/~mryder/aect_96.html.

Scouller, K.M. and Prosser, M. (1994) Students' Experiences in Studying for Multiple Choice Question Examinations. *Studies in Higher Education*, Vol. 19, no. 3.

Searl, M. (1993), Campus without walls, *Times Higher Education Supplement*, 19 November, p. 6.

Smith, G.M. (1998) Working Paper for Center for Distance Learning Research. Contact <smithgm @wpo.auhs.edu>.

Swanton, O (1998) Trouble in Paradise?, *Guardian Higher*, April 14.

Turoff, M. (1995) Designing a Virtual Classroom, International Conference on Computer Assisted Instruction, National Chiao Tung University, Hsinchu, Taiwan, March 7-10.

Twigg, C. A., and Oblinger, D. G. (1996) The Virtual University, report from a joint Educom/IBM Roundtable, Washington, D.C., Educom, November 5-6.

Vetter, R. J. and Severance C. (1997) Web-Based Education Experiences, *IEEE Computer*, May, Volume 30, No 5, p. 139-141.

Vygotski, L.S. *Mind in Society* (1978) (eds. Cole, M. John Steiner, V. Scribner, S., Souberman, E.) Harvard: London.

Endnote

1. Evaluative material may be found within the U.K. Teaching and Learning with Technology Project (e.g., Draper et al.). The authors have been granted access to a great deal of (admittedly) anecdotal case analysis within education technology conferences in the United States, e.g., EDUCOM 98, Orlando, Florida and the 8th and 9th National Conferences on Innovative Teaching and Learning, Jacksonville, Florida.

CHAPTER 11

Multimedia Development: Working Without a Script

Stephen T. Anderson, Sr.
Associate Professor

University of South Carolina—Sumter
Sumter, SC

INTRODUCTION

The multimedia craze has started to calm down, and the use of multimedia inside and outside the classroom, although it has increased, has not proliferated as widely as some thought it might a few years ago. While some classes are prime candidates for inclusion of multimedia support materials, the time necessary to transverse the learning curve and master the multimedia development tools has proven too formidable for many of us. Some attempted to "do what they could," and then their excitement was eventually replaced by the reality of time constraints. Many abandoned their early love affair with multimedia, as they did not experience the return on the investment for which they had hoped. Over the last 15 years, I have utilized multimedia software increasingly to supplement materials in my computer courses.

I will discuss the approach I chose to design and produce numerous multimedia projects, as well as give examples of those projects subject to the limitations of the printed word. I will identify what was successful and what was not as successful, and attempt to identify the reasons. I will discuss hardware and software issues and identify what proved most successful over the last few years. I will also discuss the implications of the time commitment necessary to create effective multimedia presentations, the impact of these creative and often scholarly efforts, and how administrators and other faculty members view them, especially in relation to promotion and tenure.

Multimedia Project Design

Early projects were limited to rather static slide shows for use in support of class-room activities. (Notice how I cleverly avoided the word "lecture.") The mid- to late eighties' software was severely limited by today's standards, and many were constrained to the printing of handouts or acetate transparencies, often in black and white, and most certainly without the animation, video, and digital imaging possible today. Concurrently, I tried the first "packaged" slide shows that were being marketed with textbooks of the time. I found that while the computerized slide shows had some advantages, they were mostly boring and there was certainly no sense of "ownership" on the part of this faculty member. I always felt like a substitute teacher who had been given someone else's slides and had been told to walk into someone else's classroom and "wing it." At best, it was boring. At worst, it was detrimental to student learning. As slow as ink plotters were, they at least resulted in color images that I had produced personally, allowing the feeling of ownership and commitment necessary for me to present the material effectively.

In the early nineties, I attended a pre-conference workshop on a product called "Compel." I was impressed with its power, ease of use, and its flexibility in producing multimedia projects. It also did not have the exorbitant cost associated with some of the early graphics packages, nor did it seem to have the steep learning curve of many authoring packages of that era. During the same conference, I noticed that around 10-20% of the titles of presentations related directly to multimedia concepts (the figure is between 25-40% at a typical conference today). At that same conference, I discovered an article that compared, in *Consumer Report* style, ten multimedia powerhouse software packages, including Compel, Astound, PowerPoint, WordPerfect Presentation, and Harvard Graphics, as well as some other multimedia packages.

Astound happened to receive that article's highest rating, barely edging out Compel, but clearly those two dominated the others in terms of power, price, and ease of use. I chose Astound, and I have not regretted it for an instant. While I do not want this to become an unpaid advertisement for any one product, I need to say that Astound did fulfill my three most important purchase criteria: ease of use; power; and price. I knew that while I might enjoy the greater power and flexibility of true authoring multimedia software packages, my faculty would never get excited and join me if the learning curve was formidable, as with the authoring packages of the day (and still somewhat today).

As the hardware became available and affordable to support projection of images and sound, our efforts expanded to include more than just in-class support. Projects included library support, in the form of a virtual tour, and a tutorial on the Internet, which could be "run" on one of the multimedia workstations in the library. New students were helped by a "survival computer skills tutorial," which spanned terminology, the basics of word processing, e-mail, and Internet use on our campus. Currently, we are starting to develop multimedia promotional materials to be used during business and community expositions and to be distributed via CD-ROM to high school counselors and prospective students. We are also developing cross-disciplinary tutorials on "Writing a Research Paper," as well as multimedia anatomy and biology exercises to support lab experiences. While many of these are available on our network, we are also working on dynamic HTML Web sites that will provide Internet access to much of the same material.

HARDWARE AND SOFTWARE

The good news has been the fantastic growth in power and more affordable prices of projection equipment, computer hardware, and peripheral equipment, which support multimedia development. Fast Pentium II-class computers have now become the replacement norm on many campuses. Multimedia projection equipment has become more common on all campuses. On many campuses, it is almost "expected" that those utilizing multimedia should have the equipment made available to them. Conferences almost invariably offer multimedia equipment to those presenters requesting it, while many already come with a high-lumen projector with speakers, as well as their powerhouse notebook systems. Storage of large files is no longer prohibitive, with low-cost gigabytes of hard-disk storage and inexpensive, transportable megabytes of removable media storage. Digital imaging is now commonplace, with inexpensive scanners starting from under $100, digital cameras starting from under $200, and high-end digital still cameras in the mid-$100 range. Full-motion video capture is easier than ever, but digital motion cameras are still a bit pricey for the novice.

Software has been improved to keep up with the competition and to take advantage of the other opportunities available to multimedia authors. Screen capture is a keystroke away and image editing has become user-friendly, even for the novice. Linear slide shows and simple transitions are an icon away in most of the popular multimedia presentation software. Sound creation and editing have been simplified so that it is not necessary to be a MIDI expert to create effective background sounds as well as perfectly timed narration. All in all, things are much improved for the person entering into or expanding their skills in multimedia development.

There is one cost that has improved but will never be totally contained, and that is *time*. Even though the software I have chosen is GUI oriented and push-button user friendly, there is still a time commitment that will never be eliminated in the development process. Luckily, more opportunities exist for workshops at conferences to help professionals take the fast track in the early learning curve. As more are developed, more faculty members feel the urge to come aboard.

One of my goals was to choose software that would allow faculty to be developers without requiring a change in discipline. I believe that is now possible. The vehicle I have used, Astound, allows more control over individual objects in the show and a more precise, timeline approach to event timing. I have included a screen capture of an Astound session with the timeline control dialogue box shown along with a few explanatory inserts about its components. Note that sound, animations, text transitions, video clips, etc. are all visible on this powerful control device, yet a simple "drag" of an event's duration line is all it takes to radically alter the timing and the transitions of events in the presentation.

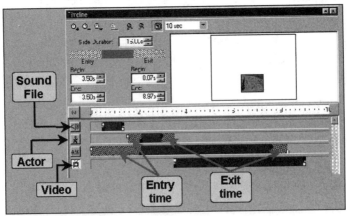

Figure 11.1

It is the timeline that differentiates Astound from PowerPoint most dramatically, plus its total flexibility and control over every single object, including multiple, simultaneous animation.

Another change in the most recent Astound (version 5.0) is the ability to save multimedia presentations as dynamic HTML Web pages. It is now possible to create animations, including object transitions and animations, at the click of the mouse. No scripting is required. No knowledge of dynamic (or static, for that matter) HTML is required. Of course, as is common to most WYSIWYG Web editors, the HTML it generates is sometimes confusing to the "old timer" HTML programmer, and it utilizes tables extensively to position objects precisely on the screen as they are positioned in the regular "slide show." While PowerPoint inserts slides as GIF images in a Web page, Astound generates a complete static Web page, saving the individual objects as graphic images on a Web page with a tiled background.

The dynamic Web pages are much more constrained, since animations must be defined in exact relation to the size of the screen on which the user is viewing the Web site. The important thing is that a novice can create an animated Web page in minutes with little or no knowledge of HTML, scripting, or any other specialized skills beyond the usual skills necessary to create a multimedia presentation. This assumes that the user has already begun transversing the learning curve of multimedia presentations.

TIME COMMITMENT

Early in the learning process, there was a significant time commitment invested in becoming functional in "the basics." I attended a workshop that allowed me to learn the basics of one software package. Even though I later chose a different package, my startup time was greatly reduced. I did choose to use my own classes to build my skills. I estimate that I spent 10 hours per week my first year learning what I could about multimedia development. I considered it a hobby as well as a professional opportunity. Early in the process, it actually was viewed more as a hobby than a creative endeavor by my faculty and administration. They never explicitly said that I was wasting my time, neither did they reward my efforts with recognition in the form of annual performance reviews. I pressed on, believing that eventually the efforts would result in scholarly presentations and publications and, at the same time, improve the learning environment inside and outside of my classroom.

In two years, I began to share my newfound knowledge with colleagues by offering conference workshops of the exact same type I attended two years before. While these workshops required significantly more time and effort than many of the papers presented at similar conferences, the recognition was minimal in the early years. I was not doing them for professional recognition; I was doing them to help others get a start on the learning curve the same way I was helped earlier.

Eventually, as the workshops became more frequent, and then became "invited" workshops, it became apparent that they were, in fact, a creative and somewhat scholarly activity worthy of at least as much recognition as a presentation and proceedings publication. The best part was that there was a lot of inter-disciplinary interest at these workshops, and I recognized that I could now start to introduce these opportunities to my non-computer faculty. The workshops have now expanded to include K-16 faculty in the multi-county service area of my campus. They are soon going to be sponsored by a local utility company so K-16 teachers can attend tuition-free. The word is out, so to speak.

The time commitment has now stabilized and has become a routine part of my daily preparation, so it is hard to measure as a separate entity. The exception is when I take on a new project, such as the recently expanded workshops or the public relations multimedia materials I am starting to create. Also, I spent around 10 hours revising our e-mail tutorial to account for our switch to new client software. When amortized over the time this tutorial will be in use, it represents a small commitment. The good news is that the creation of multimedia presentations, especially those that are visible to the general public or useful in external workshops, has been recognized during the annual evaluation process as a productive and worthwhile effort. The paper presentations have always counted, but often the in-class materials have been dismissed as simple class preparation, not accounting for the extra time involved in revamping a delivery system.

Assessment

While I have not experimentally measured the difference between the use and non-use of multimedia materials in a classroom setting, I feel there has been a difference in student learning and enjoyment. I know for a fact that there has been a measurable difference in me. The time I spend thinking and rethinking the material we cover in my classes has increased since I now try to creatively package it each time we cover it. I try not to revert to "using the show from last semester" (unless there is a significant time constraint that forces that approach.) I now find I recreate nearly all topics as I revise and improve the materials. One of my early paper presentations was entitled "Multimedia Development: Recreating the Computer Literacy Class," but the sequel was entitled "Multimedia Development: Recreating the Computer Literacy Professor."

I have not found multimedia materials to be equally effective in all disciplines. In fact, I was once a mathematician and was "retread" into computer applications as more and more of my interests leaned in that direction. I have utilized spreadsheets and other tutorial software in teaching mathematics, which have served as a supplement to in-class interaction, but I have utilized multimedia tutorial software more outside of class. I do find that there is no substitute for watching and listening to someone think through a mathematical problem in order to identify their weaknesses and to help them improve their ability in the logical processes of mathematics.

THE LIKELY FUTURE

If we broaden the definition of multimedia software to include collaborative learning tools such as WebCT and WebBoard, then I see the next few years as going in a different direction. I have recently begun to reinforce the normal on-campus interactions with Web-based threaded discussions and chat capabilities that allow the virtual classroom to expand worldwide. All one needs is a Web browser and Internet access to collaboratively learn with others in the class, students/instructors from other sections of the class, even students from other institutions in similar classes. This requires less preparation before the fact and more hands-on development and interaction during the process. While the type of preparation differs in some respects, there is one common element: *time*. I have included a screen shot of WebBoard, the software we chose to increase interactive, asynchronous, and collaborative learning at the University of South Carolina Sumter. This inexpensive tool allows us to increase communication and collaboration among everyone on campus. We even have a professor who is installing a WebBoard for the AAUP and another professional organization to which he belongs.

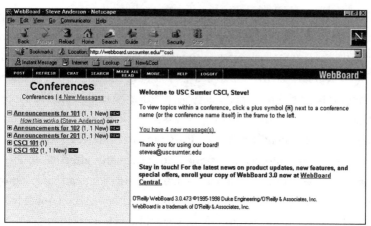

Figure 11.2

With WebBoard, we hope to collaborate more with our students as well as have them collaborate among themselves outside of class. We also hope to have our students collaborate with the students at the technical college nearby. Different sections of the same course can now discuss and chat in real time about class topics.

The learning curve of utilizing multimedia software has been considered by most to be a formidable one. WebBoard has consistently been viewed as elementary and well worth the learning effort. The "payback period" of developing multimedia materials may last longer than some are willing to wait; however, the payback period for utilizing this collaborative learning software is rather immediate. We have twelve faculty members from two neighboring institutions who, after one two-hour workshop, are ready to implement WebBoard in their classes this fall. After three years of encouragement, only three faculty members have attempted serious utilization of multimedia software in their classes. If this trend extrapolates, it suggests that the uses of collaborative learning software have a much higher probability of obtaining a critical mass than the use of previous forms of multimedia software. It certainly seems to have attracted a base of users more quickly than the latter. The use of multimedia authoring software will still remain popular with those who have already bought into it, including this author.

Although Astound allows for the creation of animated Web sites with no knowledge of dynamic HTML, the restricted bandwidth and slow download times of the typical home modem will, in my opinion, slow the growth in popularity of dynamic pages. The use of more complex techniques, such as "push technology," will remain popular and effective for those who devote full-time personnel to that endeavor, but I do not see that as a supplemental activity to be engaged in by full-time teaching faculty active in their discipline. Time will tell, but I believe collaborative Web-based software represents more of a substantial change in pedagogy than the use of other multimedia software, such as PowerPoint or Astound. I believe its impact on education will in many ways be more widespread and within a much shorter time frame.

CONCLUSION

While I believe that the utilization of multimedia techniques has changed my pedagogical style in the teaching/learning environment, I do not believe it was as widely accepted as many thought it might be. The reasons may include the perception of an inordinately long learning period, early problems with hardware availability in the classroom, dependability of the hardware, or uncertainty of payback in the teaching/learning environment. Conferences still are rich in multimedia content, and some organizations are almost exclusively devoted to multimedia, such as the Association of Applied Interactive Multimedia.

Clearly, the future of education has been drastically altered by the advent of powerful and inexpensive personal computers. The exact form that change will take is unclear. Ten to 15 years ago few, if any, of the faculty at the University of South Carolina Sumter would have guessed the profound impact of the Internet browser and the World Wide Web on education. I believe we will utilize these new technologies not to eliminate the traditional classroom, but to supplement it. The ability to communicate with students and to have them collaborate with others at great distance on an asynchronous basis will likely change the way my "office hours" are viewed. For the first time this semester, I will post virtual office hours on the Web in the form of open chat sessions held on evenings, which I will announce via our WebBoard threaded discussion conferencing system. Not bad for someone who asked, "What is the World Wide Web?" at a conference in 1991!

CHAPTER 12

Distance Education: Augmentation to Traditional Classroom

Shahram Amiri

Vice-President for Information Technology

Stetson University
Deland, FL

INTRODUCTION

Today's universities face unprecedented challenges in attempting to provide effective learning environments for their students. The goal of education should be more than purely the acquisition of information (Robson 10). In order to achieve this goal, universities must work to teach students real-life experience, diversity, and how to become active learners. Among the most obvious of these challenges are dramatic changes in student demographics, scarcity of material resources, and overcrowding of classrooms. If these difficulties are to be countered and overcome, technology will be required to play a variety of increasingly crucial and more visible roles in higher education. One of the most important of these roles will involve the adoption of technology as a primary system for instructional delivery through both distance education and a traditional classroom environment.

CONCERNS OF TODAY'S UNIVERSITIES

Public universities have long been addressing overcrowded classrooms, but private schools are just now beginning to feel the effects. According to the 1996 Statistical Abstract of the United States, the number of high school students in the United States is expected to increase by 16% between 1996 and 2006. During the same period, college enrollment is expected to increase by 14%. Universities

163

constantly struggle to meet this demand, but building more classrooms may no longer be the primary relief. A new effective solution must be sought.

Another concern for universities is the increased diversity of the student population, not only in race and culture, but also in age. Many universities are now predominantly non-white, and other schools are following the same trends. Hispanic and Asian Americans are the fastest growing college populations. Meanwhile, the average age of the college student is becoming older, while the traditional 18- to 25-year-old population is decreasing. In 1996, 41% of college students were over 25 years old. Most of these adults have full-time careers and are continuing education for career promotion and improved job performance. They need a flexible program for continued education, requiring universities to find different means of education.

TECHNOLOGY AND EDUCATION

An enhancement to education that has been extremely successful in recent years is technology. Technology is changing the traditional role that education has played and is producing more independent ways of seeking knowledge. Technology in the traditional classroom setting allows students the opportunities to gain information that would be extremely difficult to obtain otherwise. Technology has even expanded outside the classroom setting to allow students to partake in a new form of "distance education."

Technology can only enhance the present classroom; an historical example of this enhancement was the invention of the printing press. Before its invention, people were able to read and write, but not much reading and writing occurred. When the printing press was finally introduced, books were a means of preserving knowledge without relying on memory. Students were able to analyze and reanalyze the information that was presented to them without having to write everything the teacher had said. If present trends continue, it is very likely that the impact of computers on education will be proportionate to that of the printing press on writing (Withrow 61). As one can see today, the emergence of books in education did not totally replace the lecture style — it proved to be only a supplement to the classroom (Berge & Collins 1). This is true with the use of computer technology in education. As new technology increases, old technologies are broadened but not totally replaced.

An alternative to a purely traditional teaching program is to use technology in collaboration with the traditional classroom. Technology opens doors to a more productive, hands-on environment that is limited only by the technological components that serve the institution. At some schools, students are able to see time-elapsed, complex scientific experiments that, due to time constraints and other factors, they would otherwise be unable to experience. When viewing from their computers, students can alter conditions to see how the change affects the outcome of the experiment. By doing this, the usual passive form of traditional learning is changed, allowing the student to actively construct their knowledge with less guidance from the teacher. Students become more enthusiastic and motivated when they are involved in the instruction and have a choice of what, how, and when to learn. The information begins to have meaning, and the student then has an incentive to retain and use this information (Berge & Collins p. 1).

Technology redefines the classroom by breaking down barriers that exist within the confined space of a college campus. It brings to the classroom the opportunity to involve outside scholars who would normally be unavailable due to travel costs and time schedules. Not only can students meet outside speakers, but technology also offers students the chance to meet other students from different social, cultural, and economic backgrounds using Web-based instruction. This builds what Lorraine Sherry calls "new social skills," because the student learns to communicate and collaborate with widely dispersed people (Sherry 9). This is especially important because education should "transmit the culture values and lessons of the past to [the] current generation; prepare our children for the world in which they will live (Molnar 63)." Increased cross-cultural communication and peer interaction can achieve both of these goals.

The incorporation of computers and other technology forces students and educators to become technologically literate if they are to use the system properly. These skills are needed to compete in the constantly changing work environment. Incorporating these skills into the traditional curriculum only benefits those who aspire to do more with their education. Students can gain a competitive advantage in the job market through acquired computer knowledge. Students are offered a chance to learn in a new and exciting technological environment. They increase their skills by becoming capable of adapting to rapidly changing work environments.

The involvement of technology in the classroom has proved to be an effective tool; however, it still does not combat many of the difficulties that universities are facing. Technology can play only a small role in challenging the diversity of students, and it does little to combat the problems associated with overcrowded classrooms. This is where distance learning enters the picture allowing technology to play an even larger role in education.

DISTANCE LEARNING

Distance learning has been considered as a solution to help eliminate some of the problems in higher education. Distance learning cannot only help universities deal with the increasing number of students who attend each year, but can also provide students with a more diverse education. Many universities have already embraced distance learning and its advantages. By next fall, 85% of all universities with enrollments between 3,000 and 10,000 students, and 90% of institutions with over 10,000 students, expect to be offering at least a few courses that involve distance education (Blumenstyk p. 1). Distance learning includes all the benefits that increased technology can provide and more.

Distance learning allows more students to participate in education by helping universities combat the problems associated with overcrowding. It has allowed for the expansion of enrollment without building new universities or campuses. Many universities are now investing in remote access facilities rather than building more classrooms. This allows students to be spread across the campus and be instructed in a manner similar to a classroom environment. Distance learning also helps meet the needs of non-traditional students who have a hard time finding classes at the right time or do not have the time to spend commuting. Distance education, along with improved technology within education, can have a tremendous impact on the future of education. Nevertheless, the implementation of technology into the curriculum must be thoroughly planned. Distance learning cannot solve every institutional problem; clear goals and objectives for implementation need to be determined.

OBSTACLES OF DISTANCE LEARNING

Despite the seemingly endless benefits of implementing the program, distance learning cannot completely replace traditional learning. A study on a totally computer-based distance education course was conducted at higher education institutions in Indiana, and participants were surveyed about their concerns about the program. The teachers came from six universities in Indiana and had experience teaching distance education courses. The students were in the same class and were predominately married, over the age of 30, and had occupations. Over 80% of the students indicated that they had a positive educational experience.

The Indiana investigation made important discoveries regarding computer-based distance education used without a traditional classroom setting. The experimenters learned that one of the highest concerns for participants in the study was the teacher's and the student's competency with the technology and software with which they will interact in the program. This shows the need for intense campus training about computer-based instruction and increased after-hours availability of technical support on campus. The study summarizes the problem by stating, "Students cannot participate in computer-based distance education when they do not have adequate skills to use computer applications where the computer is the only means of communication and delivery for teaching and learning" (Nasseh 3).

The learner/facilitator relationship has to be defined in ways to insure proper interaction. Over 98% of the students in the study had concerns about communication between student and teacher. This type of concern may have had a major influence on the experiences of both the instructors and the students, and distance learning raises many questions in student/teacher interaction (Nasseh 5). Students still need the human interaction and experience as part of their learning process. They need to be given some of the structure that is associated with traditional learning. Even the most detailed of syllabi and course outlines cannot replace face-to-face interaction.

The interaction within distance learning is the key to success, and interaction involves thorough planning. It includes learners receiving timely feedback regarding course assignments, exams, and projects, as well as establishing contact with the instructor. Besides interaction between students and faculty, there must be a system of increased interaction among students. Since the views of other students are an important part of the classroom, this interaction would be an important criterion to the effectiveness of learning. This could be achieved with interactive databases, interactive audio and videoconferences, and e-mail. Learners must also be provided with a detailed course syllabus and chapter outlines. Also, learners need to have a working knowledge of the use of all technologies involved in the course (Distance Education). If all of these factors are in place, the process of implementing technology into traditional curricula will be more effective.

Spontaneous decision-making is difficult in a distance-learning environment. With any educational facility, effective learning is dependent on "real world examples relating to the course material." Often, an instructor may find that a point needs to be made concerning a certain aspect of the information being presented and follow tangents. This is an effective teaching tool in a traditional setting but may not be effective in distance learning due to the amount of preparation required. Considering the cost of a distance-learning facility, it is crucial for the professor to stay on the topic to maximize value to the students. To follow a tangent can become time consuming and costly to the institution, because more time may be required to ensure complete coverage of the course material. Lack of spontaneity is one of the limitations of distance learning and can be improved through increased integration. However, there are some restrictions placed on the kind of courses to which distance learning should be applied.

The question of what and how information is to be distributed among classrooms continues to be a great concern among universities. However, many times universities implement distance education programs with little regard to how they will be used. Large amounts of money are invested to equip classrooms with state-of-the-art technology, but no plan is made to use it. These educators just assume that if they have the equipment, the plans will magically appear. Unfortunately, when the educators realize that plans are needed, the equipment is already outdated and their investment is obsolete.

Even when adequate plans are made before a large investment of capital, problems arise regarding which subjects will benefit from the implementation of technology. Information can be presented in many ways to the average class, whether through case studies, lectures, group discussions, visual aids, or hands-on experiences. The problem exists in what information is most appropriate to each of these ways of presentation. Students cannot learn about computers simply by watching. Computer courses are more effective when taught in a kinesthetic environment. Likewise, classes in culture and philosophy require large amounts of human interaction for which computers cannot fully substitute (Massey & Zemsky p. 2). However, concept classes, such as journalism, history, and political science, could use distance education quite effectively.

Recently, there has been quite a push for information technology along with distance learning. However, there are still many concerns about the implementation of these new ideas. It seems that telecommunications services are experiencing a mixed reception among educators (Withrow 59). Many information experts who have advocated information technology in the past agree that there has to be a balance between traditional and distance education. They realize that this is the best method for the introduction and effectiveness of distance learning. They stress that distance learning should not be an independent and isolated form of learning.

IMPLEMENTATION OF TECHNOLOGY INTO THE CLASSROOM

Incorporating technology into traditional curriculum is a benefit to anyone who decides to further his/her education. However, along with distance education and technology, there are many responsibilities. The implementation of technology into the curriculum must be thoroughly planned. Technology cannot solve every institutional problem. Its implementation must have clear goals and objectives established to be successful. We cannot continue to advocate the use of information technology unless we are willing to take the time to design an educational program that embraces the technology so that it can enhance the present classroom.

A major problem with learning today is the tendency to confuse information with learning (Bork 70). It is important to remember that "We cannot become enamored by all the available technologies without dealing with the educational purpose we are trying to fulfill with the technology" (Sherry p. 1). Instead of just incorporating technology into education, we need to improve the approach we take to distance learning to maximize the value it adds to education. The use of technology in education has greatly influenced the way students participate in education. Technology allows for a new integrated form of learning (Berge and Collins p. 1).

CONCLUSION

Education has changed with the introduction of technology and will continue to change. It has grown from "an orderly world of disciplines and courses to an infosphere in which technology is increasingly important" (Molnar 68). Clearly, the future of education will see a major change with the expansion of technology. However, that change depends on our willingness to adopt technology into educational institutions and the manner in which we administer it.

References

Berge, Zane and Mauri Collins. "Computer-Mediated Communication and the Online Classroom in Distance Learning." *Computer-Mediated Communication Magazine*. April 1, 1995. http: Hsunsite.unc.edu/cmc/mag/1 995/apr/berge.html.

Blumenstyk, Goldie. "Small Colleges Lag in Distance Learning, Education Department Survey Finds." *The Chronicle of Higher Education*. October 7, 1997. http:Hchronicle.com/ che-data/news,dur/dailarch.dir/971 O.dir/97100703.htm.

Bork, Alfred. "The Future of Computers and Learning." *Technological Horizons in Education—THE Journal*, June 1997.

Barry, Willis. "Distance Education at a Glance." Engineering Outreach: University of Idaho. http:www.uidaho.edu/evo/distl.html.

Massey, William F. and Robert Zemsky. "Using Information Technology to Enhance Academic Productivity." Educom: Interuniversity Communications Council, Inc., 1995.

Molnar, Andrew R. "Computers in Education: A Brief History." *Technological Horizons in Education—THE Journal*, June 1997. Ebsco Host Academy-Online database under Educom Review.

Nasseh,Bizan. "Computer-based Distance Education in Higher Education Institutions in Indiana." Ball State University: http://www.ind.net/IPSE/CONCLUD.html.

Robson, Joan. "Some Outcomes of Learning through Teleconferencing." Australian Catholic University: http:Hwww.usq.edu.au/electpub/e-jist/robson.htm.

Sherry, L. "Issues in Distance Learning." *International Journal of Distance Education*, 1996. http://www.cudenver.edu/public/education/ edschool/issues.html#issues.

Withrow, Frank B. "Technology in Education and the Next Twenty-Five Years." *Technological Horizons in Education—THE Journal*, June 1997.

CHAPTER 13

An Encounter with Proteus: The Transformational Impact of Distance Learning

Eugene J. Monaco
Project Administrative Officer
Professional Development Program

Kary Jablonka
Lecturer
School of Social Welfare

Rebecca Stanley
Senior Education Specialist
Professional Development Program

University at Albany (SUNY)
Albany, NY

Constant, accelerating, and often unpredictable, change presents significant challenges to maintaining workforce competencies and proficiencies. The ability of organizations to quickly and effectively respond to changes in the operating environment; incorporate new knowledge and skills; and understand and implement new policies and programs, directly depends on the quality of workforce development practices. The education, research, and community service mission of the university speak directly to that challenge.

The educational enterprise that aims at preparing individuals for entry into the professional workforce and at maintaining the competence and proficiency of the existing workforce is engaged in a mythic encounter with Proteus—the shape shifter of Greek lore. The son of Poseidon, Proteus is a prophetic sea divinity. Usually found near the Island of Pharos, where he herds Poseidon's seals, Proteus emerges from his cave each day to sleep in the sun. People wishing to know the future must approach while he sleeps and grab hold of him. However, the instant he is caught Proteus assumes a horrifying array of monstrous forms to scare the person into letting go and, thus, avoid prophesying. If held fast despite his best efforts to frighten, he resumes his usual form of an old man and tells the future. Such is the nature of the encounter between the university and distance learning.

Distance learning is one of the most rapidly evolving dimensions of the changing environment. It may well be that the university as we know it won't survive as the emerging future takes on an increasingly non-traditional character. Recounting the many prodigious efforts that threatened to make faculty, administrators, students, and outside partners let go of the future, this chapter explores the transformational impact of integrating a distance-learning dimension into a well-established, multi-million dollar, university-based continuing professional education program that is based on working partnerships with state government agencies. This encounter with Proteus—the shape shifter who knew all things past, present, and future—is forcing a redefinition of teaching and learning, and of relationships with external partners and the organization itself.

Beginning with a brief overview of the history of the university's involvement in continuing professional education and its experience in establishing and maintaining working partnerships, this chapter describes the initial involvement and continuing engagement with distance learning (encompassing satellite-based teleconferencing, compressed video, point-to-point desktop videoconferencing, and the WWW) in the delivery of undergraduate, graduate, and non-credit programming. Most importantly, it will identify and discuss the implications for three critical domains of the educational enterprise: teaching and learning; external relations; and the higher education institution itself.

THE PROFESSIONAL DEVELOPMENT PROGRAM

Functioning as the continuing professional education and training sector of the University at Albany's Nelson A. Rockefeller College of Public Affairs and Policy, the Professional Development Program (PDP) administers over $15 million/year in statewide education and training activities. A leader in providing professional development services to state and local government since 1976, PDP has an established capacity to effectively and efficiently design, implement, and operate a full array of educational services, including needs assessment, curriculum development, and program delivery for a broad range of public sector client agencies. Activity during that time largely consisted of traditional, classroom-based undergraduate/graduate education and non-credit platform training.

BACKGROUND

With the installation of a statewide satellite network (SUNYSat) throughout the 64-campus State University of New York as well as City University of New York, PDP incorporated the production and delivery of one-way video/two-way audio satellite-based programming in 1988. This decision stemmed from the reported effectiveness of television as an instructional medium (Moore and Kearsley 1996; Russell 1992). This was soon followed by the use of electronic bulletin boards and CD-ROM technology to deliver content to geographically dispersed target audiences.

Continuing to expand the use of new learning technologies, PDP now utilizes compressed video, Web and Intranet-based instruction, as well as satellite and traditional delivery modes to offer graduate education and non-credit workshops. The addition of new delivery vehicles continues to carry implications for the nature of the educational enterprise, the academic institution, and relationships with external partners. The first graduate telecourse, offered during the 1996 fall semester, served as a crucible that began to crystallize these issues. Lessons learned have guided subsequent efforts in delivering three additional graduate telecourses: Overview of Substance Abuse and Human Behavior and the Social Environment I and II.

The pioneering effort, Introduction to Child Welfare, a three-credit graduate social welfare course, was delivered via Ku-band satellite to over 30 sites across the state of New York using the traditional one, three-hour class/week for 13 weeks format. Not realizing the consequences of adopting the established academic paradigm, the instructor was committed to producing and delivering three hours of live television per week for 13 weeks. Approaching the course as if it were a traditional on-campus offering further compounded the challenge with the assignment of three 15-page papers. Over 300 students produced in excess of 13,000 pages of material that had to be read, critiqued, and graded. This was perhaps the easiest challenge to meet by assigning a graduate assistant to help manage and grade assignments.

Since many studies have either found no significant difference between learning outcomes (Hayes and Dillon 1992; Simpson, Pugh, and Parchman 1992) or improved results (Martin and Rainey 1993), production began in earnest during summer 1994. The closely related production issues that had to be immediately confronted were: 1) finding a balance between "good television" and sound graduate education; and 2) bridging the gap between the instructor and students. To be effective, the medium demands more than "talking heads" reflecting the traditional lecture approach. This raised the notion of production values—those elements that add to what the instructor says and create additional venues for students to interact with the material. Several features were incorporated into each class to address these concerns.

THE IMPLEMENTATION OF DISTANCE LEARNING

To help anchor content in the practice environment, interviews with child welfare caseworkers were videotaped and edited. These short segments were used by the instructor to illustrate that particular week's topical area and as a point of departure for discussion. Another feature was a weekly "In the News" portion that reviewed relevant news clippings submitted by students from receive-sites around the state. This helped students understand that the issues are systemic in nature—deeply embedded in the service delivery system, rather than the result of the idiosyncrasies of geography or personality.

Student-instructor interaction is an essential element in ameliorating what is referred to as transactional distance (Moore 1990). A function of how much interaction and to what extent it is structurally incorporated into course design (Moore 1990; Saba and Shearer 1994), transactional distance is the relationship between interaction and structure. Therefore, additional course support was provided by a Web site developed as an extension of class time to help create the sense of a virtual classroom. Included was the syllabus, graphics used each week, a discussion group with weekly discussion topics, and hot links to relevant Web sites. In an effort to stimulate their use of the Web, students were invited to submit Web sites that were then featured as part of each week's class with a "Webcrawler of the Week" award bestowed on the student submitting the selected Web site. Other production elements included using one or more guest panelists, video clips to introduce issues and frame discussion, and small-group activities to be done at each receive-site during a break.

As well as efforts to enhance the in-class connection between students and the instructor, "office hours" were established with a toll-free number with voice mail back-up, e-mail, and a fax line. This proved to be a challenge since the advisement and support needs of the large, dispersed class were substantial and included guidance as to what "get into the literature" meant, how to write term papers, how to access materials, matriculation, and the significant question of socialization into the professional culture of social work practice and values.

Five areas have emerged as critical dimensions of successful instructional design and course delivery: 1) faculty training and support; 2) helping learners learn; 3) infrastructure/technology access; 4) working within budget; and 5) value-added nature of the virtual classroom. First, support and technical assistance is needed to assist faculty in adapting to the virtual classroom. Deprived of the opportunities afforded for spontaneous feedback and interaction, the transition to the distance-learning environment requires an orientation to the available technology and how to use it. There are also the larger questions of pedagogy versus andragogy, what makes for effective instruction, and whether the integrity of the intellectual enterprise may be compromised.

CONCERNS OF DISTANCE LEARNING

Although always a consideration, helping learners learn is an issue magnified in the virtual classroom. Efforts need to be made to orient students not just to the graduate experience but to how it will be different in a distance-learning environment. A related aspect to this is the sort of access students have to the needed equipment. While there may be substantial capacity on the production and instructional side, students may lack regular access to a computer or the Internet or be using equipment that is incompatible in some regard. From this perspective, it is the technology available to the least equipped student that influences course design.

The Financial Investment in Distance Learning

The financial dimension of distance learning is perhaps the least understood. While there is seductive power in believing that it is less expensive than traditional instruction, this is not necessarily the case. Poised at the frontier of the virtual classroom, there are expenses associated with pioneering effort that must be managed as part of the price of shaping an important part of the future of education. It is interesting that distance education initiatives quickly face demands for data related to return on investment that are never applied to the traditional classroom setting. Clearly, more work is needed to determine the value-added nature of distributed learning, how best to address the question of costs and what constitutes cost, and the relative effectiveness of different teaching and learning methods. What has been learned about teaching and learning in the virtual environment is that team teaching should be used whenever possible; students should be provided with feedback on how they are doing early in the semester; a Web site is a valuable and essential element; and institutions must view the investment as building for the long-term.

Distance Learning Transforms the University

Just as teaching and learning are being transformed by distance education, so too is the university as an organization, how work is organized, and partnerships with external partners. When engaging in the encounter with Proteus, there is great temptation to cordon it off in a separate box on the organizational chart where faculty and staff are insulated from the effects of the many faces of prophecy that must be endured. Simply put, while this may be a viable strategy for a short time, the challenge is best met by integrating distance learning into the entire organization. Ultimately, this entails a rethinking and realignment of functions and units, staff development, curriculum design, linkages with the operating environment, and the need for continuous development and promotion. The new learning organization will consist of synchronous and asynchronous learning platforms with instruction accomplished by teams of technical experts.

CONCLUSION

Despite the desire for a model or prototype to guide involvement in distance learning—in effect to "peek around the corner" as a way of reducing risk and fear—it is only in the willingness and tenacity to hold fast to Proteus that the future of the virtual classroom can be crafted. As the steam engine propelled the world from the Agricultural into the Industrial Age, the microprocessor is the steam engine of today that is launching the Information Age, and the consequences of that for the educational enterprise will be no less dramatic.

References

Hayes, J.M. and Dillon, C. (1992). Distance Education: Learning Outcomes, Interaction, and Attitudes. *Journal of Continuing Education for Library and Information Science*, 33(1), 35-45.

Martin, E.D. and Rainey, L. (1993). Student Achievement and Attitude. *The American Journal of Distance Education*, 7(1), 54-61.

Moore, M.G. (1990). Recent Contributions to the Theory of Distance Education. *Open Learning*, 5(3), 10-15.

Moore, M.G. and Kearsley, G. (1996). *Distance Education: A Systems View*. New York: Wadsworth Publishing Company.

Russell, T. (1992). Television's Indelible Impact on Distance Education: Where We Should Have Learned from Comparative Research. *Research in Distance Education*. October, 2-4.

Saba, F. and Shearer, R. (1994). Verifying Key Theoretical Concepts in a Dynamic Model of Distance Education. *The American Journal of Distance Education*, 8(1), 36-59.

Simpson, H., Pugh, H.L. and Parchman, S.W. (1992). *The Use of Videoteletraining to Deliver Hands-On Training: Concept, Test, and Evaluation*. San Diego: Navy Personnel Research and Development Center. (NPRDC-TN-92-14).

CHAPTER 14

Developing Web Assignments in English as a Second Language and Multicultural Education

Inez A. Heath
Associate Professor Social Science and Multicultural Education
Valdosta State University
Valdosta, GA

Cheryl J. Serrano
Associate Professor College of Education
Assistant to the Dean, College of Education
Lynn University
Boca Raton, FL

INTRODUCTION

The continued improvements in instructional software, as well as advances in telecommunications such as the Internet (Net), electronic mail, and distance learning, are changing the way educators think about teaching and learning. The field of education is responding to the fact that the world is becoming more global. Events that occur in the United States affect other parts of the world and events in other countries impact the United States. Thus, the implications for education are abundant in response to these conditions. The need for educators to gain a global perspective and to increase students' global awareness is imperative.

With a long tradition as experimenters in new technologies for instruction, foreign language and English as a second language (ESL) teachers are especially enthusiastic about the recent breakthroughs in computer-assisted language learning (CALL). The emphasis on a natural approach in teaching language has been effectual in shifting the emphasis of foreign and ESL instruction from traditional methods that teach language in isolation to a more academically integrated and interactive approach that emphasizes the development of language through interpersonal learning experiences. As the Internet and the World Wide Web (WWW) become more accessible, teachers are realizing the value of this resource for enhancing the development of English language skills within the academic curriculum in multilingual classrooms.

The Internet's appeal as a tool for enriching education lies in its limitless possibilities for interactive activities, independent thinking, developing language skills through comprehending and using authentic (natural) language, and interpreting information. With a few clicks of the mouse and ample time for exploration, teachers are able to design activities that integrate content from across the curriculum into their daily lesson plans.

In his presidential initiative on technology in education, President Clinton (U.S. Department of Education 1998) emphasizes that "by the year 2000, 60% of the new jobs in America will require advanced technological skills." He strongly encourages parents, teachers, and leading CEOs to work together on a national mission to equip schools with the right technology to ensure that children will have the tools for success in the future. His four-pillar challenge includes: the accessibility of modern computers for every student; classrooms that are connected to each other and to the world; the integration of educational software in the curriculum; and teachers who are ready to use and teach technology.

The challenge for teachers working with linguistically diverse students is to integrate academic content with classroom activities that present opportunities to expand and reinforce language skills based on the student's proficiency level, also keeping in mind the student's age, interests, cultural background, and prior educational experiences. This is a lot to contemplate when one considers the vast amount of information already available on the Internet.

Two questions come forth: how to guide teachers in defining and evaluating Internet resources for English language learners (ELL); and how to prepare teachers to train their students to become discriminating and intelligent decision-makers when evaluating the credibility and accuracy of all this information. Meloni (1998) writes that in spite of the debate over the value of the Internet for ESL/EFL (English as a Foreign Language) teachers and students, there appears to be consensus and support for this powerful resource as being valuable in increasing student motivation and natural communicative competence in the new language. Additionally, few would argue against its value in contributing to students' global awareness.

Compared with textbooks and teacher-centered instruction, the Internet offers students learning experiences that are personally and locally relevant, along with current information, including graphics and sound that are colorful and "alive." More and more instructional designers are developing Internet sites for language learners and are paying special attention to improving the interactive aspects that include graphics and sound, making text more comprehensible. As these capabilities improve, the potential for creating instructional materials that are far more interesting and effective will expand. Unlike software programs on CD-ROM, the Internet allows instructional designers to update their sites periodically, taking into account current events and paying attention to pop-culture and linguistic variations that are characteristic of our media-driven language.

Our failure as a nation to provide adequate opportunities for language minority students to succeed in school has led to a high dropout rate among many of these students. Garcia (1997) writes, "On almost every indicator, non-Whites and Hispanic families, children and at-risk students are likely to fall into the lowest quartile or indicators of 'well-being'...."

Among the indicators included is educational achievement. The result has been an increase in minority-language adults who lack the skills to function successfully in workplaces where technological skills are required.

A recent City of New York "Technology for Learning" study (U.S. Department of Education 1998) found that use of computers can improve learning and educational opportunities for at-risk students. The researchers concluded that significant academic improvements were especially evident in reading when computers were provided in the homes of at-risk middle school students. The President's Technology Initiative and studies such as these have been instrumental in leading efforts to end the "digital divide" through programs that focus on cutting costs and expanding computer accessibility to the at-risk population.

One example, the E-rate, is a special discount that enables at-risk schools to receive telecommunication service, such as the Internet and WWW with a 20-90% sliding-scale discount, is based on the number of students in attendance who are eligible for free lunches. In spite of its success, there are members in

Congress who oppose this effort. It is, therefore, important for those involved in education to stay up-to-date on worthwhile legislation.

Recent initiatives to eliminate bilingual education programs and decreases in federal funding for the training of bilingual/ESL professionals have also compounded the situation, making it necessary for school districts to train teachers to infuse ESL methods and strategies across the curriculum while using English as the medium for instruction. Although many school districts have ignored this issue, problems remain as the number of linguistic minority students continues to increase across the nation. The need for teachers who are adequately prepared to work with linguistically diverse students in multilingual classrooms is still critical.

TEACHER TRAINING PROGRAMS: HOW CAN THEY ASSIST IN THE TASK?

While we cannot expect all teachers to have a strong background in second language teaching, educators need to investigate using educational technology as a resource. Courses in teacher-education programs should provide pre-service teachers with opportunities to discover the many resources available through technology. There are many strategies for integrating the WWW into the teacher education core curriculum.

Assignments that allow pre-service teachers to research information and analyze its value and potential for use with their future students will increase their confidence and motivate them to explore a multitude of resources. In turn, they will be better prepared to meet the needs of the diverse learners. At the same time, they will be exposed to language-learning principles and practices that facilitate the second language acquisition process for ELLs.

Such assignments should also require discussion of the language learning process, including understanding the developmental stages of second-language acquisition and how specific Web sites may be beneficial for designing instruction. Browsing the Net to find sites that can be used directly for instruction is also another important assignment that provides opportunities to individualize instruction.

The prospect of enhancing instruction using the computer may seem overwhelming and intimidating, especially for those veteran teachers who lack experience and training in using computers in the classroom as an instructional tool. However, with only a few hours of browsing the Net, many of these formerly reluctant teachers become enthusiastic users of this exciting resource. Thus, introductory assignments related to the WWW should emphasize an open-ended, hands-on, and anxiety-free experience that will encourage the user to become an explorer in this maze of possibilities.

Green (1997) suggests that the Internet has added a new dimension that is especially valuable for language learners. In her "journeys" through the Internet, she is continuously locating new sites that are very appropriate and meaningful for language learners. The Internet is a live source that is always evolving. Teachers today must be willing to align themselves with the technological forces that will soon impact everyone worldwide. Prospective teachers are responsible for becoming constructive critics of these Web sites. It is especially important for educational methods courses to integrate technology and to provide guidelines for constructive assessment of the various Web sites, focusing attention on a variety of pedagogical as well as current academic issues.

THOUGHTS ON TECHNOLOGY AND SECOND LANGUAGE LEARNING

An important consideration for all teachers working with ELLs is the concept of language proficiency level. Generalizations are often made regarding an individual's language proficiency level based on a given set of criteria; however, it is important to remember that the individual learner is unique. His/her language learning is based on numerous variables that distinguish him/her from other learners who may be considered to be at the same level. Krashen (1982) emphasizes that we acquire language by "going for meaning." In the "Input" hypothesis, he suggests that language is acquired when we provide the learner with input + 1—1 being the additional comprehensible information, or input, that allows the learner to expand his/her knowledge base in the new language because it is beyond the learner's current stage of second-language development.

According to Krashen (1982), communication is successful between the learner and teacher when the learner has been provided with appropriate input + 1 that is manageable. That is, the learner is provided with adequate authentic language that is understandable. The input does not deliberately attempt to aim at one particular aspect of language, but rather allows students to draw upon their existing inter-language. This experience facilitates communication. Additionally, language is best learned through content, content that is relevant and meaningful to the learner. This has implications for CALL since content presented through the WWW is authentic and natural. It provides i + 1 when the content presented via the Web site is meaningful, comprehensible, and relevant to the learner. In the process of acquiring the language, each individual relies on personal experiences and background knowledge while also using a variety of linguistic skills. Comprehension is a result of this process. Thus, language acquisition and content knowledge are enhanced simultaneously.

Cummins (1998) suggests that the more target language text learners read and comprehend, the more of the target language they learn. Cummins asks, "If the research on extensive reading is as effective as indicated, then why is it not used more in second language teaching?" He recommends that with the advancement of technology, there are many possibilities for encouraging ELLs to read and acquire language using technology. Cummins states that computer-assisted language learning ... harnesses the vast amount of authentic naturally occurring language in text and transforms this language into comprehensible input that fuels the language acquisition process.

EXPLORING WEB RESOURCES FOR ENGLISH LANGUAGE LEARNERS

There are many resources that classroom teachers working with multilingual students can adapt for use in their K-12 classrooms. For students at the beginning level of proficiency, for example, the teacher may focus instruction on Web sites that assist in developing basic aural/oral, grammar, vocabulary, and syntax. It is important to emphasize that when the Internet is used with other materials, the curriculum is automatically enhanced.

Teachers working with elementary-level children should also consider Web resources that are colorful, interactive, and basic. At this level, the use of thematic units may be enhanced with materials found on the WWW. For example, in developing a thematic unit on monsters with second-grade ELLs, Green (1997) and her students found two favorite Web sites from which she copied stories on monsters to enhance the unit. Using the WWW as a resource brought students much more in touch with their thoughts and fears about monsters and things that they considered "scary" while also developing their literacy skills.

AN INTRODUCTORY ASSIGNMENT
FOR EDUCATION STUDENTS

The responsibility of staying current on new Web sites and assessing new software within content specific areas is an important and ongoing task of today's teachers' instructional development and planning. Students in education programs, whether at the undergraduate or graduate level, need to approach this task with confidence and knowledge since the field of educational technology is constantly evolving. An appropriate assignment for pre-service and in-service teachers to prepare them for use of the Internet and the WWW is an open-ended, hands-on, stress-free strategy for novice computer users to explore the numerous resources on the Web.

The task involves evaluating sites on the Web that support curricular goals. The objective is to develop a rubric or set of standards for selecting resources from the Internet, aligned with the school's goals, students' interests, their cultural and linguistic needs, and their levels of English language proficiency. Upon completing the task, students meet in small groups to discuss the Web sites they evaluated and to share the independently prepared rubric with each other. This is followed by a whole-class, teacher-led discussion, during which the group collaboratively develops a list of criteria for a generic Web site evaluation form.

DIALOGUE JOURNALS

Meloni (1998) recommends the use of dialogue journals as an exceptional strategy to improve fluency in writing for second-language students. Teachers are able to involve students at various levels of proficiency and grade levels. Students write on topics of their own choice and send their entry to their teacher. The teacher responds to the student, expanding on the topic in adapted discourse, with the goal of inviting further response from the student. There are several benefits from this informal writing assignment. The second-language students have an opportunity to use language and to expand their linguistic knowledge through dialogue with a more fluent English speaker. The teacher is able to assess linguistic weaknesses and design instruction to meet student needs.

Dialogue journal interaction can also be arranged in a variety of ways. In a middle school, for instance, the teacher had students involved in learning about history using cross-age dialogue journals. The students researched a topic and wrote to senior citizens who were living in a retirement community. This was a successful project because the teacher met and interviewed the senior citizens to learn about their past experiences prior to selecting the topics of study.

Group journals may also involve two or more groups of students responding to each other on issues and ideas that are of special concern to the two groups involved. Suggested topics for dialogue journal assignments include political campaigns, moral and social issues, planning special events, sports, and music.

KEYPALS

Similar to pen pals, keypals are pen pals who live in other states or countries and use e-mail to communicate and receive responses instantaneously from other students. Meloni (1998) notes that this is especially effective with intermediate level ESL students. There are several Web sites available for matching students based on age, interests, and other factors.

EXCHANGING IDEAS AND INDIVIDUALIZING INSTRUCTION

The Web is not only a tremendously effective means for disseminating instructional materials; it also provides opportunities for student collaboration and innovation in learning, according to Li and Hart (1996). While its use may seem limited to students at an intermediate proficiency level, teachers working with students at beginning stages of proficiency are encouraged to investigate sites such as Ex-Change. Ex-Change is a new ESL Web magazine that advocates those teaching ESL to interact with teachers like themselves through e-mail for the purpose of editing and revising instructional materials to meet students' individual needs. Li and Hart also state that there are more Web sites currently for ESL learners than for teachers, and there is an abundance of Web sites under construction for ESL students, so-called "works in progress."

Having updated information on these Web sites is important for teachers when developing activities that integrate technology.

COMPARATIVE CULTURES: LEARNING ABOUT THE UNITED STATES AND OTHERS THROUGH THE WEB

Another activity proposed by Green (1997) uses the Internet as a resource for introducing newcomers who will be attending high school to the United States. Her goal of developing a comparative perspective is achieved through assigning students to learn about their own respective cultures using the Internet. Students are asked to compare and contrast similarities and differences based on what they know about their own culture and what they have learned about North American culture. Through virtual map tours and resources in their native languages, students have opportunities to expand their linguistic ability in a non-intimidating way.

Additionally, students research recipes and songs from their countries, which also encourages language development. Students may consult bilingual dictionaries, parents, and cultural informants as resources in translating the materials. The final experience is a virtual tour of the White House and the opportunity to send an e-mail message to the president.

WORLD GEOGRAPHY THROUGH INDIGENOUS CULTURES

Encouraging teachers to learn about the cultures and indigenous groups that exist and influence life in many of their students' countries is another assignment that may be enhanced through the use of the Internet. This activity emphasizes the diversity among indigenous people throughout the world—their lifestyles, traditions, languages, and the often remote and interesting places where they live. The purpose of this task is for students to learn about their communities and their cultures and to make connections between their lives and other ethnic groups around the world.

The series used in this assignment, *"Voices of Forgotten Worlds"* (Blumenfield 1993), includes CD-ROMs that feature the music indigenous to cultures around the world, with accompanying text about each culture. It has been an effective resource for encouraging comparative study of cultures and world geography among teachers working with linguistically diverse students. This assignment integrates cultural geography, incorporating maps, text, authentic music, language study, and photography. Learning about each culture is enhanced through research on the Web, which adds currency, authenticity, and real-world understanding.

Students in small groups are assigned readings that are culturally specific. Each group is required to complete an in-depth study that includes using the Web to locate supporting documentation of current information on the particular culture. The value of this task is that it allows students to explore the ancestry and tribal origins of many modern cultures and in doing so, it extends understanding of their connections to cultures around the world and throughout history.

TRUTH AND LIES ON THE WEB

An activity developed for secondary in-service and pre-service teachers by Marcowitz (1996), "Nazis on the World Wide Web," provides a unique way to assist students in evaluating the accuracy and credibility of informational sources that are on the Web. According to Marcowitz, people are exposed to a wide variety of information from numerous resources—some good, some bad, some biased, some true, and some lies.

This activity allows students to explore a highly controversial and difficult topic, Nazis, and to become "critical consumers of information"(Marcowitz 1996). This is by far one of the most valuable lessons teachers can present to students. It is a challenge for native English language speakers and an even more complex task for ELLs, since subtleties in language use can manipulate perceptions and understanding of ideas that are negatively biased and deceptive. Nevertheless, this kind of activity, while difficult, is becoming increasingly important as individuals rely more on information from the WWW.

This approach requires a great deal of planning and understanding of the issues involved. The emphasis on a problem-solving approach requires students to read a variety of perspectives on the topic of Nazis available on the Web. Students discuss the information and respond to analytical questions related to the readings. This is the first step in "critical information literacy" (Marcowitz 1996), which is one of the most important steps that classroom teachers and teacher trainers should consider as their students use the WWW for research.

RESEARCHING RESOURCES FOR ESL TEACHERS ON THE WEB

With the continual additions and changes in home pages, it is important that teachers with shared interests develop Web home pages and share highly-rated sites with peers. Stebbens (1996) provides a list of Internet/WWW sites that have value for all teachers. The Internet is an exciting experience, as information is easily accessible and relatively inexpensive. Several sites that are useful for ESL teachers are highlighted in her article, as well as funding sources in education. Also, provided is a list of general education sites. Education journals such as *Sunshine State TESOL Messenger* and *TESOL Journal* occasionally include a section on technology. The national TESOL organization has a well-established educational technology interest section. These journals offer reviews and updates on new Web sites that are specific to the TESOL field. These valuable resources are a must for teachers working with ELLs.

TECHNOLOGY-INTEGRATED INSTRUCTION WITH THE ENGLISH LANGUAGE LEARNER IN MIND

The most important aspect of Web assignments is to provide prospective teachers with a strategy to evaluate Web sites, especially in terms of their accuracy of information and their value for learners. Furthermore, one must consider how these sites will enhance learning. A guide or checklist that assists the user in determining the value of Web sites should be used as part of all assignments involving Web sites as resources for teaching. A guide, as such, has been developed to evaluate Web sites and may be modified based on the needs of the user (see Appendix). The criteria given are exclusively for use in assessing Internet sites related to ELLs. In assessing an Internet site for ESL students, the teacher should consider the following factors: the students' linguistic needs; curriculum goals in various grade-levels; academic areas; age and developmental level; prior academic experience; native language; cultural background; and specific interests.

CONCLUSION

While the idea of a technology-integrated curriculum may appear to be a far-away notion to some, it is already a reality in many schools nationwide. The push for a technology-enhanced curriculum is an ever-present challenge for all involved in education, especially when we consider the diversity among our students in the United States. There are those individuals who fear that technology will eventually replace teachers. While such fears are unfounded, teachers must nevertheless realize that their roles are changing. They are no longer the sole source of information in the classroom nor the central figure in controlling and imparting knowledge in the classroom. Rather, the new technology has expanded the teacher's role as a facilitator in the acquisition of knowledge and information, and the need for teachers to be learners. With these changes come new and exciting pedagogical opportunities, as well as responsibilities.

The technology-integrated curriculum requires well-trained decision-makers who are able to facilitate and guide the process of learning. These requirements also include the need for content area specialization, media skills that are current, and creativity. The ability to develop effective, individualized lesson plans that foster independent thinking for learners will prepare students for life-long learning in a technology-driven world.

In addition to being "technologically smart," teachers will continue to provide the human element that is essential not only in aiding the learning process, but more importantly, in shaping the psychological, social, and moral development of the individual. These elements are inherent in preserving cultural heritage and are essential to all societies. Educators face countless new challenges as they continue in their roles as advisors and mentors, providing the affective support for students to explore an interconnected and increasingly complex world.

References

Blumenfield, L. (Ed.) (1993). *Voices of Forgotten Worlds: Traditional Music of Indigenous People*. New York: Ellipsis Arts.

Cummins, J. (1998). E-lective language learning: Design of a computer-assisted text-based ESL/EFL learning system. *TESOL Journal* 7 (3), 18-21.

Garcia, E. (1997, March). The education of Hispanics in early childhood: Of roots and wings. *Young Children*, March, 1997, Vol. 52, No. 3, 5-14.

Green, L.C. (1997). Cruising the Web with English language learners. *IDRA Newsletter*, 24 (5), 9-13.

Krashen, K. (1982). *Principles and practices of second language acquisition.* New York: Pergamon Institute of English Press.

Li, RC. and Hart, R.S. (1996). What can the World Wide Web offer ESL teachers? *TESOL Journal,* 6 (2), 5-10.

Marcowitz, D. (1996). *Nazis on the World Wide Web.* Unpublished manuscript, Loyola College, Baltimore, Maryland.

Meloni, C. (1998). The Internet in the classroom: A valuable tool and resource for ESL/EFL teachers. *ESL Magazine,* 1 (1), 1-16.

Regional B/2 Title I Technical Assistance Center (1995-96 First Quarter).Transforming teaching, learning through computer technology. *Focus on Education,* 1-7. [Online newsletter]. Available at: http://www.ed.gov/Technology/.

Stebbens, C. (1996, fall). Using the Internet for ESOL research. *Sunshine State TESOL Messenger,* Vol. 2, No.1, 9.

U.S. Department of Education: Office of Intergovernmental and Interagency Affairs (1998, July, August). Bridging the technology gap in schools. *Community Update,* No. 59, Washington, DC: Author.

APPENDIX
WEB SITE EVALUATION FORM

General Information:

Web site name: _____

URL Address: _____

Web site authors/organization/corporation/agency/institution:

_____

Search engine used to find the site: _____

Headings or topics under which this site is listed: _____

_____

Appropriate grade level: .

    Elementary ___   Middle ___   High ___   Adult Learner ___

English proficiency level:

    Beginning ___   Intermediate ___   High Intermediate ___

How would you integrate this Web site in your classroom? _____

_____

Language Development:

Briefly explain how these skills can be developed and enhanced through use of this Web site, and rate each area.

    There is evidence of exemplary vocabulary, grammar, syntax.

        low 1   2   3   4   5   high

    Related terms and concepts are highlighted.

        low 1   2   3   4   5   high

    Academic language is developed within the context.

        low 1   2   3   4   5   high

    Cultural aspects of language are evident.

        low 1   2   3   4   5   high

Opportunities for oral language development are provided.

low   1   2   3   4   5   high

Adequate opportunities to develop literacy skills are provided.

low   1   2   3   4   5   high

II.   *Home Page Information:*

Written language used on the Web site is clear and concise.

low   1   2   3   4   5   high

Advertising is tastefully done and age-appropriate for students.

low   1   2   3   4   5   high

Information about affiliation of the group represented on the Home Page, and its mission, is evident.

low   1   2   3   4   5   high

Objectives related to the development of this Web site are clear.

low   1   2   3   4   5   high

Address, phone numbers, and names of individuals involved with Web site are available.

low   1   2   3   4   5   high

Information on Web site reflects accuracy in research and knowledge about subject(s) discussed.

low   1   2   3   4   5   high

III.   *Links and Hypertext:*

Links work and are reliable: Links are well organized, and progress logically. Frustration with links is minimized.

low   1   2   3   4   5   high

Information on most links is useful and relevant.

low   1   2   3   4   5   high

The user has the opportunity to communicate with others through a chat line, bulletin board, or e-mail.

low   1   2   3   4   5   high

Briefly discuss and give an overall rating for any enhancing features, such as audio, music, speech, special effects, etc., which are included on this site.

low   1   2   3   4   5   high

IV. *Graphics*:

Graphics are clear and relevant. They support (not overpower) the Home Page design

low   1   2   3   4   5   high

Graphics support information about Web site. They provide an accurate representation of the people, places, and mission of the authors.

low   1   2   3   4   5   high

Graphics enhance text and provide helpful information for English language learners.

low   1   2   3   4   5   high

V. *Reliability and Stability*:

(This part of the evaluation requires periodic checking of the site over several weeks.)

Web site is well established.

low   1   2   3   4   5   high

The site is periodically updated and current.

low   1   2   3   4   5   high

Periodic changes improve navigability of site.

low   1   2   3   4   5   high

Instructional Value and Appropriateness of Web site:

Evaluating sections: In each section, add scores, and divide by number of items.

Evaluating Entire Web site:

Add the raw scores for each section and divide by 22, or by the number of all items to which you responded in your evaluation.

5=Excellent   4=Very good   3=Average.

Any score below 3 indicates that this Web site is not recommended for your students.

CHAPTER 15

Virtual Languages: An Innovative Approach to Teaching EFL/ESL (English as a Foreign Language) on the World Wide Web

Cheryl J. Serrano
Associate Professor, College of Education
Assistant to the Dean, College of Education

Lynn University
Boca Raton, FL

Randall L. Alford
Associate Dean, College of Science and Liberal Arts
Chair, Division of Languages and Linguistics

Florida Institute of Technology
Melbourne, FL

INTRODUCTION

Increasingly, we hear more and more about the popularity of the Internet and its myriad uses. The Internet, the network of networks built by the Department of Defense for its own purposes, has now become a symbol of the 1990s. With access to computers, modems, interactive videos, and CD-ROMS, individuals now have entry to a world beyond their immediate environment: the capabilities of interacting with the "rich and famous"; finding answers to questions from experts almost instantaneously; staying in touch with both family and friends from within one's own country; and getting acquainted with citizens of the world. In the educational arena the Internet (NET) enables students and teachers to work outside the classroom, the school, and even their country in new and exciting learning venues, some real and some virtual. "This places school building design and curriculum materials in a new light," according to Gonzalez (1997).

How wide spread is this interest in cyberspace? *The Wall Street Journal* (Weber 1998) reports that almost 62 million people in the United States use it, representing 30% of United States residents who are 16 years and older. The NET reaches into the homes of almost a quarter of all North Americans. Membership in the popular America Online, Inc. is greater than the combined readership of *The Wall Street Journal*, *The New York Times*, and *USA Today*.

The Wall Street Journal (Weber 1998) also published a unique profile of Internet aficionados. The typical "Webber" is wealthier, with an average income of $55,000 a year; better educated, with 43% holding college degrees; dominated by males (58%); and is an average of 37 years old. This profile is an indication that the availability of computer technology in the home will widen the gap in educational opportunity between those families unable financially to invest in such costly equipment and those families who are affluent.

Outside the United States as one would expect, most of the World Wide Web (WWW or Web) users are from developed countries. Li and Hart (1996) examined Web page usage from Li's "English as a Second Language" (ESL) Home Page to gain insight into the nature of the Web ESL community and its needs. Access to their six-section Web page reached as high as 2,000 hits per day. They concluded that the listening and speaking section experienced the most hits, accounting for 36% of total accesses. About 46% of the identifiable accesses came from the United States, and the rest came from more than 40 other countries, with Korea, Japan, Canada, Australia, Brazil, Italy, Germany, Taiwan, Hong Kong, and Israel identified as significant users.

Educational professionals are becoming increasingly more interested in computer technology and its application to teaching and learning, as learners are more diverse than ever before, enrolling in United States all-English classrooms from a multitude of linguistically and culturally diverse backgrounds and academic needs. The population of children for whom English is a second language is projected to reach 3.5 million by the year 2000. This population, excluding children of undocumented workers, is expected to approach 6 million by the year 2020 (Pallas, Natriello, and McDill, as cited in Faltis 1997). The challenge of meeting the diverse needs of all learners is one that keeps educators constantly searching for innovative strategies. Classroom teachers, as facilitators of the learning process, need to find, select, and offer information to their students in a variety of ways that integrate the acquisition of academic language and content.

"By the year 2000, 60% of the new jobs in America will require advanced technological skills" (Region B/2 Title I Technical Assistance Center 1996), which indicates that the time to initiate learning through computer technology is now. The United States Department of Education is looking at the Internet to help prepare today's students for tomorrow's technical world. Businesses and communities have been invited to collaborate with educators to ensure that all students in the United States will be able to log-on to the superhighway's vast storehouse of knowledge. The developed world has moved from the Industrial Age to the Information Age, in which economic activity and growth are based less on the input of more labor and more on the exchange and interpretation of information and the development of knowledge. Therefore, the ability to read, write, and communicate effectively over computer networks will be essential for success in almost every sphere of life (Warschauer and Healey 1998).

Such a national mission has been the impetus for the development of four goals by President Clinton in his Educational Technology Initiative: (a) teachers will have training to assist students in learning to use computers and the super-highway; (b) all teachers will have computers in the classroom; (c) every class-room will be connected to the information superhighway; and (d) effective and engaging software and online resources will be a part of the curriculum that teachers and students can use (Region B/2 Title I Technical Assistance Center 1996). The research clearly indicates that the infusion of technology across the curriculum acts as a catalyst for both gifted students and those who previously may have been unmotivated or may have difficulty with traditional teaching methods and are at risk of dropping out of school. Teachers who are involved with the integration of technology into classroom activities also appear to have higher expectations of their students. Technology seems to empower students to engage in all aspects of learning tasks; to develop higher-order thinking visualization and literacy skills; and to be familiar with creative uses of computer technology. Clifford, of the Defense Language Institute, states that, "Technology will not replace teachers; teachers who use technology will replace teachers who do not" (as cited in Meloni 1998). Educators cannot afford to ignore the potential of the Internet, and yet it should not be used only because it simply exists. It should be accessed because it enhances learning experiences and offers something unique that students cannot get in routine classroom activities.

Meloni (1998), in her article "The Internet in the Classroom," states that the use of the Internet increases student motivation; uses authentic language; makes us globally aware while simultaneously increasing global understanding; and is environmentally friendly, allowing for the conservation of natural resources. These are all good, sound reasons to maximize this marvelous educational resource with our students, in particular linguistically and culturally diverse students who are in the process of acquiring ESL and a new culture. There is evidence from an informal survey conducted by Teachers of English to Speakers of Other Languages (TESOL) that a growing number of university ESL programs are using information technology to promote programs and provide student support, such as e-mail communication, and student orientation to the campus and community, along with a number of other functions.

COMPUTERS, LEARNING, AND LANGUAGE LEARNING: JUST HOW FAR HAVE WE COME?

As technology becomes integrated into the teaching/learning process, the role of the classroom teacher changes noticeably. Classroom teachers become facilitators who assist students in constructing their own understandings and capabilities in carrying out tasks on the computer. The shift from lecture and recitation, which often still occurs in secondary classrooms, to coaching automatically supports a *constructivist* approach to learning. The introduction of this third party, the computer, encourages the teacher to play the role of a coach (Collins 1991). Interestingly, the classroom teachers are drawn to those students who are in need of assistance and students who may be academically weaker. This shift from working with better students during whole-class instruction to individual instruction has been documented in several school sites (Collins 1991).

In addition, another benefit of using computers is the reported dramatic increase in students' engagement in long-term activities and projects. Students become invested in the activities they perform on the computer. Using computers entails active learning, which often uses visual media to enhance the learning process. The integration of both verbal and visual forms facilitates comprehension and, in turn, promotes language learning.

Consequently, society's prevailing view of education is slowly, inadvertently changing instruction as computers and electronic networks potentially provide instant access to the world's accumulated knowledge, in both verbal and visual form (Collins 1991). The World Wide Web, a single electronic communications network, is making widespread computer-based instruction a visual reality. There are many reasons that the Web has become a popular environment for implementing computer-based instruction: the ease with which documents can be created; worldwide accessibility; multimedia capabilities; and interactive functions (Li and Hart 1996). The nature of school tasks/assignments has begun to reflect what has already been occurring in the corporate world. The effective use of technology can support and advance key elements of school restructuring and reform, including student-centered classrooms, student exploration and real-world problem solving, interactive instruction, heterogeneous grouping of students, performance-based assessment, multi-disciplinary learning, and teacher as facilitator. These elements reflect the underlying philosophy of constructivism.

Computer-assisted language learning (CALL) has an extensive history of more than 30 years in the United States. Interestingly, there have been three main stages of computer usage that have corresponded to teaching practices and research and a particular level of technology during each stage. The "behavioristic CALL" implemented in the 1960s and 1970s featured repetitive language drills, referred to as drill and practice.

Eventually, as new findings emerged from extensive research being conducted in the area of second language acquisition, the next stage, "communicative CALL" emerged in the late 1970s and the early 1980s. Supporters of this philosophy stressed that computer-based activities should focus more on using language functions than on the forms themselves. In other words, the teaching of grammatical structures would be implicit rather than explicit. Fluency took precedence over accuracy in the target language. Second-language learners were encouraged to generate original language utterances rather than manipulate prefabricated language that did not reflect their current stage of second language acquisition/interlanguage. The rationale corresponded to cognitive theories and theories of second language acquisition that emphasized that learning was a process of discovery, expression, and development (Warschauer and Healey 1998).

In recent years, there has been a broader reassessment of the communicative language teaching theory and practice to a more social or socio-cognitive view that emphasizes language use in authentic social contexts. The goal is to integrate language learners into authentic environments while infusing the various skills of language learning and use. There are three main approaches being used: task-based; project-based; and content-based. The three approaches involve the language learner in using language for specific purposes. In the third stage of computer usage, Warschauer and Healey (1998) describe this relatively new perspective on technology and language learning, "integrative CALL," as "a perspective which seeks both to integrate various skills (e.g., listening, speaking, reading, and writing) and also integrate technology more fully into the language learning process." Students experiencing integrative CALL have opportunities to learn to use a variety of technological tools in an ongoing process of acquiring and learning a second language. At the same time, students must be exposed to a variety of materials that allow and encourage them "to explore and be creators of language" when the goal is communication in the target language (Warschauer and Healey 1998).

DESIGNING WEB LEARNING MATERIALS FOR SECOND/FOREIGN LANGUAGE LEARNERS

The responsibilities of curriculum and materials developers are of utmost importance. These specialists must have a clear understanding of new technologies, interpret that potential in the light of what they know about the needs within the profession, and develop and produce programs that can use those technologies appropriately and effectively for a targeted population of learners, in this case, English language learners. Their obligation is similar to that of the teachers' obligation to students: To find connections between technologies and the teaching of English to speakers of other languages and academic content (Murison-Bowie 1993). Li and Hart (1996) remind developers that any TESOL courseware on the Web must be considered "work in progress" since technology is evolving rapidly and continually reshaping the uses of computers in language learning.

There is tremendous potential for developing language, literacy, and conceptual abilities by providing learners with visual and audio support by integrating computers and other interactive technologies into the curriculum. According to Hunt (1993), who extensively reviewed innovative computer-based programs that exemplified the principles of good instructional design and the best practices for promoting second language learning, there are specific characteristics of materials that may be classified as exemplary multimedia products. These characteristics are as follows:

- Flexibility—The programs can be effectively used across a range of grades and levels of proficiency.

- Thematic presentation—The programs introduce and reinforce vocabulary and syntax within a rich contextual framework.

- Appropriate content—The programs for adolescents and adults deal with relevant, current issues, and those meant for elementary children are age-appropriate.

- Multiple modalities—The programs provide many opportunities for students to listen, speak, read, and write.

- Open-ended questions or writing prompts—They encourage students to take risks with language by expressing themselves with creative, unique responses.

- Natural interaction—The program's online, and suggested off-line, activities provide students with opportunities to communicate with each other in a natural, meaningful way.

- Mixed media—In addition to the laser disc, computer, and/or CD-ROM resources, the programs offer complimentary audiotapes or print material.

- Extensive system guides—These guides provide information on the use of the components, with numerous suggestions for activities in lesson adaptation, extension, and evaluation.

- In-service—The publishers offer assistance through staff development videotapes or on-site workshops.

The work of Li and Hart (1996) has focused on creating multimedia learning environments for intermediate-level ESL students because they concluded that second language learners at this proficiency level can gain most immediately from Web-based courseware. The multimedia capabilities and interactive functions make the WWW an ideal environment for carrying on computer-based instruction for an individual with intermediate proficiency in English.

The Development of Virtual Languages

The Web-based distance learning program entitled, Virtual Languages, is designed to promote language learning for high-beginner and intermediate to advanced English Language Learners. Virtual Languages is a developing member of Educom's Instructional Management Systems (IMS), whose primary goal is to increase professionalism in the development of distance learning courses worldwide. The Virtual Languages online English language course uses WebCT, a content tool, which facilitates the creation of sophisticated Web-based educational environments. This tool allows educators to design the appearance of course pages; it provides a set of educational tools that can easily be incorporated into any course; and it provides a set of administrative tools that assist the educator in the task of course administration.

For these reasons, WebCT has been selected in designing the Virtual Languages program (Goldberg and Salari 1997). Through its unique design as an online course, any of the following learner environments may be selected: The learner's own home, a computer lab, a self-access room in a private company, or a public area in a school.

The Virtual Language General English Course program, which is thematic in nature, incorporates authentic readings in the target language across numerous content areas, such as Aviation and Space, Life in the United States, Environment, Health, Sports, Travel, Entertainment, Biographies, and Business. Additionally, Virtual Languages offers specialized English language development courses for pilots and air-traffic controllers, business professionals, and hotel and restaurant personnel. It also customizes programs for other service areas.

The goal of Virtual Language is to facilitate the acquisition of both social and academic English through a variety of innovative activities related to interesting, relevant, and age-appropriate readings that introduce students to the language they will encounter in the English-speaking world. Lessons are designed for upper-elementary children, teenagers, and adults at three proficiency levels. An English language proficiency assessment instrument, the Adapted Test of English Proficiency Level (ATEPEL), is available to adolescents and adults who are interested in more accurately determining their proficiency prior to selecting a particular course level. Students will be exposed to English from a variety of American dialects and speakers of varying age groups. According to Crandall (as cited in Sokolik 1998) authentic readings challenge English language learners to increase their understanding of both content and culture. Authentic readings instill motivation within the learner to continue developing the target language, which is a critical variable when considering the length of time it takes to develop communicative competence in a new language.

The lessons include built-in support, which Cummins (1998) describes and promotes for facilitating comprehension for presenting target language text. The Virtual Languages program supports combined text, sound, colorful graphics, and glossed vocabulary in context; a glossary; practice formats; moderated chat rooms; bulletin boards; and e-mail for course participants. The developers are seeking to infuse all of the language skills with technology, which is reflective of integrative CALL. By incorporating bulletin boards, chat rooms and e-mail, language learners will have opportunities to use English for authentic purposes: discussions around selected themes with instructors, guest content-area specialists, and classmates—potentially from continents around the world. A calendar of events will organize and post dates and times for discussion topics related to the readings and will identify discussion group leaders.

English Language Learners will be invited to interact through modified language input, which Faltis (1997) names "comprehensible invite." It is similar to comprehensible input in that it is adjusted to the learner's needs and is a little beyond the learner's current stage of proficiency. Comprehensible invite encourages the learner to interact using the creative construction of knowledge within discourses related to the topic for discussion and is similar to comprehensible input in that it is understandable to the learner. This is an opportunity for learners to negotiate for meaning with the teacher, discussion group leader, or peers. Faltis (1997) states that the "negotiation of meaning encourages them to try out new language as well as to notice the gap between their existing language and the language they need to more clearly and precisely express themselves."

Comprehensible invite is automatically supplied again during the process of negotiation. According the Warschauer (1998), the language used in computer-assisted discussion has been shown to be more lexically and syntactically complex than in face-to-face, discussion which indicates that this activity is a good strategy for English language learners who are motivated to make progress at a faster rate. Discussion group leaders and instructors will be able to e-mail students in order to provide individual feedback when requested.

CONCLUSION

As previously stated, coursework on the Web must be considered "work in progress" (Li and Hart 1996) since the capabilities in technology are continuously advancing and doing so at different rates around the world. One of the distinct advantages Virtual Languages has in designing instruction on the Web is the potential to make curricular changes as needed based on feedback from English language learners or the needs of corporate-specific clients. The plan is to continue to expand its curriculum by adding lessons based on identified needs, to further develop its listening and speaking activities, and to continue to integrate built-in support to facilitate comprehension of the target language. Another long-term goal is to include interactive videos related to the lesson themes when the Web end-user has the technological capacity for such innovation.

The developers of Virtual Languages are collaboratively attempting to develop an English language instructional program. This program will exemplify the characteristics of good instructional design and the best, current applied linguistic practices for promoting the acquisition and learning of English. The developers are doing this in an effort to prepare their students "to function in a networked society...," where "English is likely to remain the lingua franca of the new global society..." (Warschauer and Healey 1998).

References

Collins, A. (1991). The role of computer technology in restructuring schools. *Phi Delta Kappan*, 73 (1), 28-36.

Cummins. J. (1998). e-Lective language learning: design of a computer-assisted text-based ESL/EFL learning system. *TESOL Journal*, 7 (3), 18-22.

Faltis, C. J. (1997). *Join fostering adapting teaching for the multilingual classroom* (2nd ed.). Upper Saddle River, New Jersey: Prentice-Hall.

Gonzalez, J. M. (1997, May). Technology in education: Time to face the monster. *IDRA Newsletter*, 24 (5), 8-9, 14.

Goldberg, M. W. and Salari, S. (1997). An update on WebCT (World Wide Web Course Tools)— A tool for the creation of sophisticated Web-based learning environments.

Proceedings of NAUWeb'97—Current Practices in Web-Based Course Development, 1-11 [online]. Available at: http://homebrew. cs.ubc.ca/webct/papers/nauweb/full-paper.html.

Hunt, N. (1993). A review of advanced technologies for L2 learning. *TESOL Journal*, autumn (1), Vol. 3, No. 1, 8-9.

Li, R. C. and Hart, R. S. (1996). What can the World Wide Web offer ESL teachers? *TESOL Journal*, 6 (2), 5-10.

Meloni, C. (1998, January/February). The internet in the classroom. *ESL Magazine*, 10-16.

Murison-Bowie, S. (1993). TESOL technology: Imposter or opportunity. *TESOL Journal*, autumn (1), Vol. 3, No. 1, 6-8.

Region B/2 Title I Technical Assistance Center. (1995-1996 First Quarter). Transforming teaching, learning through computer technology. *Focus on Education*,1-7. [online newsletter]. Available at: www.ed.gov/Technology/.

Sokolik, M. E. (1998, June/July). Authentic readings for all learners. *TESOL Matters*, Vol. 8, No. 3, 15.

Warschauer, M. and Healey, D. (1998). *Computers and language learning: An overview* [online]. Available at: www.///.hawaii.ed/markw.

Weber, T. E. (1998, September 16). Who, what, where: Putting the internet in perspective. *The Wall Street Journal*, Web interest section.

CHAPTER 16

Using an Internet–Based Distance Learning Model to Teach Introductory Finance

Sharon H. Garrison
Professor of Finance

Daniel J. Borgia
Assistant Professor of Finance

Florida Gulf Coast University
Fort Myers, FL

ABSTRACT

This chapter focuses on the development of an Internet-based distance learning model for teaching the introductory finance course in the finance department at America's newest institution of higher education, Florida Gulf Coast University, which opened in August 1997. An important component of the mission of FGCU is the incorporation of technology and the needs of the regional community into curriculum design. Keeping these charges in mind, the finance department paid particular attention to building a curriculum by consensus by inviting more than 80 community business leaders to participate in focus groups to provide input about the kinds of skills they would value in FGCU graduates. Because of its commitment to the university's mission, the department felt it was important to develop a separate Internet-based course as an alternative to the traditional in-class introductory finance course. In this Internet-based course students are required to participate in a "boot camp" for the first few weeks. During this part of the class, students are given instruction on only the most complex aspects of the course. After this initial period, the course is completely Web-based in design. The Web page for this course, located at http://www.tmag.com/sgarrison/courses/fin3240d/index.html, contains a variety of pedagogical materials to help students learn and comprehend course content. It contains the class syllabus, learning objectives, solved problems, class lectures, and a "Web Site of the Day." The Web page was designed to be aesthetically appealing but not flashy so that it would load quickly and be easy to navigate.

USING AN INTERNET-BASED DISTANCE LEARNING MODEL TO TEACH INTRODUCTORY FINANCE

Background

Teaching finance in a university is never an easy task. Finance relies heavily on complex theoretical concepts and applications that must somehow be effectively conveyed to students. Most applications are technique-oriented. Many students experience considerable difficulty in mastering both the concepts and techniques, often because of inadequate math backgrounds. Another difficulty is that the field of finance is continuously evolving in response to changes in technology and market conditions. At Florida Gulf Coast University (FGCU), there are even greater challenges to be overcome. Students at FGCU come from a variety of backgrounds. Some are quite skilled in math and in accounting principles, and some are lacking in these areas.

Florida Gulf Coast University, the tenth university in the Florida State University system, opened its doors in August 1997. The university's mission encourages innovative teaching techniques and utilization of technology. The finance department in the College of Business supported this mission. The marriage of technology and finance is continually strengthened as new information technologies are continually integrated into the financial markets.

ADVANTAGES AND DISADVANTAGES OF DISTANCE LEARNING

One component of the university's mission was to try to deliver distance-learning courses to 25% of the students enrolled at the university. The finance department accepted this challenge, yet realized there would be significant barriers to overcome to effectively teach finance in a distance setting. As the use and role of distance education was implemented and expanded at FGCU, however, it was imperative that we examine both the prospects and problems implied prior to embarking on this journey. This meant it was important to identify the advantages and disadvantages of a distance-learning delivery system.

The first obvious advantage of distance learning is that courses otherwise unavailable on-site because of low enrollments or instructor availability can be made available to interested students. Secondly, there is a natural constituency of potential students for whom actual class attendance is difficult or impossible. Examples include people who live far from campus, people with physical disabilities, and people whose work schedules do not allow them to attend class regularly.

Potential hurdles to the development of distance education include a lack of trained faculty, unfamiliarity of both students and faculty with required technology, and a potential reduction in faculty-student interaction and communication.

OTHER UNIVERSITIES' EXPERIENCES WITH DISTANCE LEARNING

To help design our course, we chose to focus our research efforts on the mathematics literature because, like finance, it requires intensive problem solving and analysis on the part of students.

In 1991, the North Carolina School of Science and Mathematics (NCSSM) (Wilson 1997) initiated an early distance-learning program. Established in 1980, the NCSSM was the nation's first statewide residential public high school for eleventh and twelfth grade students with potential talent in science and mathematics. Administrators of the NCSSM distance-learning program found that to perform well in a distance learning course, a student must be highly motivated, self-disciplined, and able to work independently without constant supervision. Furthermore, teachers should possess certain qualities, including flexibility, to cope with lapses in technology; the ability to develop assignments designed to foster both individual and cooperative behaviors conducive to learning; and the ability to think linearly in order to organize the material clearly for presentation.

In the spring semester of 1998, the department of mathematics at the University of Colorado offered two courses over the Internet as the first stage in its distance-learning program (Abrams 1998). A distinguishing feature of the program was the incorporation of traditional classroom elements into the presentations. This was done because of the belief that the two most important components of teaching mathematics are visual transmission (using some sort of chalkboard) and verbal transmission (the voice/words of the instructor to describe and explain the material). The University of Colorado created the Internet portion of the course using an audio and "whiteboard" conferencing system called Rendezvous by VisualTek Solutions, Inc. In addition, they selected Microsoft's NetShow application to deliver the audio feed of lectures over the Internet.

In a study to determine the effectiveness of using distance technology to teach mathematics, researchers used qualitative research methodology to explore the perceptions of students and teachers participating in an interactive, collaborative, satellite-based mathematics course (Larson and Bruning 1996). Results showed that the distance-learning format gave teachers access to more resources, was useful for underachieving students, and was an effective way to implement national curriculum and instruction standards.

Northern Virginia Community College (NVCC) developed a distance-learning program for the mathematics, science, and engineering courses required to complete an Associate of Science degree in engineering (Sener 1996). The results indicated that students achieved completion and grade distribution rates comparable to students attending on-campus offerings of the same courses.

In a study comparing the scores of on-campus and off-campus students taking a common final examination in similar graduate education courses at Nova Southeastern University, researchers found that the 36 off-campus students had higher average scores on the 30 common examination questions than their 25 on-campus counterparts (McFarland 1996).

A mathematics professor at California State University at Northridge divided his statistics class in half, teaching one group traditionally in class and another in an online version of the course using a World Wide Web site, electronic mail, and an electronic chat room (McCollum 1997). On both midterm and final exams, the latter group performed significantly better.

THE INPUT OF LOCAL BUSINESSES

In designing courses for distance at FGCU, the faculty also sought the advice of local business leaders who met in focus groups. The focus groups were asked what skills they wanted to see in finance graduates. In addition, potential students were polled as to their preferences in course design, timing, availability of resources, etc. The finance faculty also wanted to build a sense of identity among finance majors. Some of the facts we learned were:

1. Businesses wanted to hire students with good communications skills.

2. Businesses wanted graduates to have "hands-on" financial skills.

3. Students wanted classes at convenient times and were quite interested in distance-learning modalities.

4. Students wanted courses that were comprehensive, yet carefully designed to optimize their time.

DISTANCE-LEARNING COURSES

One example of a distance-learning course designed in response to these findings is Finance 3240, the introductory, undergraduate finance course. The Web page for a sample course is located at http://www.tmag.com/sgarrison.

Finance 3240 is taught both in class and at a distance. The distance version of the class has been taught three times with good reviews from the students who completed the course. The attrition rate for the course is high (27%). However, this compares favorably to the attrition for distance courses at the university (32%). In reviewing the Finance 3240 Web page, there are a number of items to note. First, there are a number of topic overviews that give a thumbnail sketch of the topic for discussion. These are designed to explain some of the theoretical concepts in simpler terms and to add practical applications that employers desire.

There are also a number of tutorials to give students "hands-on" financial training. Some of the tutorials cover the basics of using financial spreadsheets, the basics of working with the time value of money, and how to use spreadsheets in valuing corporate securities. The course page contains the syllabus for the course as well as some general materials that we developed, such as sample exams and various tutorials. We also have a page devoted to each chapter in the course textbook.

The chapter materials are objective-driven, as learning objectives are listed for each chapter and all subsequent materials derive from those objectives. We also include a chapter outline, solved problems, and a "Web Site of the Day." In addition, we allow students to review the PowerPoint slides that we use in our classes. These are viewed using a PowerPoint animation player. We purposely do not put the actual PowerPoint slides on the page to minimize download time. Also, we have observed students in university classes printing out PowerPoint slides in the lab. This is not a good use of university resources, particularly when in our class we have the chapter outlines available for student note-taking.

Perhaps the most important feature of the page is the Feedback Form. We decided that we could not wait until the end of the semester to make course improvements. Instead, our goal was to make continuous improvement. To that end, we give students the opportunity to provide us with feedback on every lesson. Most of the time, the comments have been fairly gratifying. Occasionally, we get a good suggestion that we can use immediately to improve the course. For instance, the first semester we taught this course, we made e-mail assignments due on Fridays. A couple of students wrote us suggesting that we move the due dates to Sundays at midnight. This gave them the opportunity to work on assignments over weekends. Students forward many of our best "Web Sites of the Day" to us. By responding quickly to suggestions such as this, we feel that we are treating students as stakeholders in the course.

OTHER FINANCE COURSE WEB SITES

We frequently peruse other courses in finance. Few distance courses are currently being offered, but we try to remain in contact with other finance professors who work in a distance environment to keep abreast of new developments. See http://www.cob.ohio-state.edu/~fin/resources_education/ edcourse.htm.

CONCLUSION

Teaching finance in a distance environment is a challenging process, but input from stakeholders in the process optimized efforts. The class that we developed has been operating smoothly and we continually seek to improve it.

References

Abrams, Gene. (1998). "S.H.O.W.M.E.: Spear-heading online work in mathematics." *THE Journal* 25 (10), pp. 53-56.

Larson, Matthew R. and Robert Bruning. (1996). "Participant perceptions of a collaborative satellite-based mathematics course." *American Journal of Distance Education* 10 (1), pp. 6-22.

McCollum, Kelly. (1997). "A professor divides his class in two to test value of on-line instruction." *Chronicle of Higher Education* 43 (24), p. A23.

McFarland, Thomas W. (1996). "Results from a common final examination: A comparison between on-campus students and off-campus students." Research and Planning Report 96-17, ERIC microfiche, Document no: ED403821, 16 pages.

Sener, John. (1996). "Delivering an A.S. engineering degree program through home study distance education." ERIC microfiche, Document no: ED393493, 20 pages.

Wilson, Virginia. (1997). "Distance learning." *Journal of Secondary Gifted Education* 9 (2), pp. 89-102.

CHAPTER 17

Technology-Based Learning and Adult Learners: Lessons from Practice

Dolores Fidishun
Head Librarian

Penn State Great Valley School of
Graduate Professional Studies
Malvern, PA

INTRODUCTION

Most people who work in higher education today are very aware of the need for students to learn to use technology. In addition, more and more faculty have begun to teach using technology. As this phenomenon expands, it will be critical for us to understand the implications that these technology interfaces have for adult students. A walk across any campus, particularly in the evening when graduate classes are meeting, evidences that one can no longer assume the students we teach will be the typical 18-year-old, freshly graduated from high school. Many adults are now returning to school to complete bachelor's degrees. In addition, more and more professionals are returning to earn an advanced degree to improve their career potential or to retool to begin a new career path. Tice (1997) indicates that the number of college students over age 40 has tripled since 1970. Adults learn differently than traditional students and have different needs from institutions of higher education. This becomes especially evident when students interface with technology.

To meet the needs of adult students, colleges and universities are looking for ways to integrate technology into teaching by allowing students to complete modules or entire classes or degrees away from campus via computer. The importance of this trend becomes obvious when we look at institutions like the University of Phoenix (Fischetti et al. 1998), or more traditional institutions like

Drexel University, which now offers a Masters of Science in Information Systems in a completely asynchronous mode (Drexel University 1998). Penn State University has inaugurated a world campus that will allow students to complete a number of courses and programs from a distance (Pennsylvania State University 1998).

As the need for the use of technology in education and the rise in adult student numbers converge, it becomes necessary to understand how adults learn and how educators can facilitate adult learning when designing and offering courses via computer. The following chapter will examine adult learning concepts and discuss how they can be used in the implementation of technology-based learning. Technology-based learning is defined as any program, course or part of a course, which is offered via the computer, usually from a distant or off-campus site and is predominantly asynchronous in structure as opposed to traditional classroom-based instruction.

The examples of the development of an online library catalog tutorial and an Internet-based library tour at the Penn State Great Valley School of Graduate Professional Studies will be used to point out specific needs of adult learners that can be met by using technology.

ATTRIBUTES OF ADULT STUDENTS

Adult students have different types of goals and learn differently than traditional students. Zemke and Zemke (1991) stress that adults prefer self-directed and self-designed learning seven to one over group learning and that "the desire to control pace and start/stop time strongly affects the self-directed preference." Adults are accustomed to setting their own schedules and priorities in their lives. Zemke and Zemke (1991) also state that adults "bring a great deal of experience into class," something that those teaching adults need to acknowledge and use in their design of instruction.

Prominent adult educator Malcolm Knowles (1990) talks about adult students' readiness to learn those things that they feel they need to know. He also states that adults are life-centered or task-centered or problem-centered. They are motivated to learn things that will assist them in confronting problems that they perceive in real-life situations. Lawler (1991) explains that adults have " a pragmatic desire to immediately use or apply their knowledge or skills." Instructional methodologies that allow students to set their own pace or choose to do modules or lessons of specific value to their learning needs can most effectively engage adult learners.

In addition to the desire for a connection to real-life experiences, some adults prefer working with others to help them understand a concept. Many have learned to work in teams at their places of employment and find that methodologies that allow collaborative learning can help them to talk through concepts or track parts of a skill they may have missed as the professor explained it.

Lawler (1991) also discusses the many roles and responsibilities that adults have in their lives. She explains that, unlike younger students, education is only one priority in the life of the adult student. Families, careers, and other responsibilities may have to take a greater role in their life than a class that they need to attend. It is not uncommon for an adult student to be traveling for work on a night that they are scheduled to be in class or to have to attend a child's soccer game while trying to finish a paper. It is, therefore, important that adults have the ability to cover material outside of a structured classroom setting and do research at odd times, such as late at night or very early in the morning, when libraries may not be open or faculty may not be available. It is not unusual for an adult student to start to work on a paper at 11:00 p.m., after their children are in bed and other responsibilities have been fulfilled.

An additional trend among adults undertaking education is the fact that most attend classes voluntarily, frequently prompted by a transitional event in their lives. These transitional events can vary from a divorce or job loss to a promotion or children leaving home. Merriam (1991) evidences that the reasons people give for learning consistently correspond to their life situations. This voluntary participation means that students are eager to learn and expect to have resources available to them at all times. They will seek additional materials on a topic or complete larger amounts of work if the lessons can enhance their lives.

Finally, adult students, unlike traditional students, are conscious of the perception of how much they know. Experts in other areas of their lives, they may be disconcerted to start at the beginning as they learn a new skill or concept. As Lawler (1991) states, "Adults who are functioning in competent and successful ways in their professional and personal lives may come into classrooms with trepidation and anxiety." Thus, Lawler describes the faculty member's role in teaching adults as one that encourages "questioning, risk-taking, and the tolerance of ambiguity."

If one reflects on the needs of adults as they enter higher education, it seems evident that we need to seek new techniques and methodologies, which can more accurately fulfill the demands they make for a different type of learning environment. The use of technology-based education is one way that some educators have been able to meet these demands.

TECHNOLOGY AND ADULTS

Stephen Brookfield (1986) states that adults learn best when they feel the need to learn and when they have a sense of responsibility for what, why, and how they learn. The use of technology by both faculty and students in higher education can meet the demands of adult students by providing lessons and courses that can be adapted to these students' distinct learning requirements. As the advantages of using technology-based education with adult students are discussed, an Internet-based online catalog tutorial (Pennsylvania State University LIAS Search 1998) and library tour (Pennsylvania State University Library Tour 1998) created at Penn State Great Valley School of Graduate Professional Studies will be used as an example of the practical application of technology-based learning in an adult setting.

The Internet-based library tour and tutorial were developed at Penn State Great Valley as a response to requests from students. Great Valley is a Master-only institution that offers professional degrees in education, instructional systems, engineering, management, and MIS to students in the Philadelphia region. In excess of 95% of Great Valley students work full-time in business or education and their median age is in the early thirties.

The clientele of the school are typical adult students who evidence all of the trends listed above. After hearing requests from students for access to library instruction on an as-needed basis, the librarian created an Internet-based tour of the library and a tutorial for the Penn State LIAS (Libraries Information Access System) online catalog. In developing these instructional modules, an effort was made to use as many adult-learning principles as possible to make them effective for the adult population.

Feedback received about student requirements for library instruction indicated that a self-paced, self-directed format was important. A Web-based, interactive design of modules allows students to choose when they need to access each section based on the searching strategies they must employ to do their research or their previously acquired library skills. An individual student may need simply to review how to search by series and then return to the actual LIAS catalog to do the search. Another student may want to follow the tutorial in a more linear fashion to fully understand the logic of an online catalog. Thus, students are able to follow the path for receiving instruction that they find useful based on previous knowledge or a skill needed at that moment. The addition of examples relevant to adult students and the topics they study helps to enhance the value of the lessons.

Another reason for self-directed access is that Great Valley students, as all adult students, have many responsibilities in their lives. When talking to these students, one frequently hears stories of people accessing the online catalog from hotel rooms while traveling for business or at 1:00 in the morning after children are asleep. Access to the Internet permits an adult student to receive instruction at any time, day or night, as well as from any location, as long as one can access the Net.

Adult's voluntary participation in education leads to demands for greater services and quicker responses to questions. Logistics and funds do not permit the library to be open 24 hours a day every day of the year, but technology-based modules allow students to receive instruction in catalog search techniques or library policy at times convenient for them. Coupled with the ability to submit interlibrary loan requests via the Web, this service allows students at Penn State Great Valley to do much of their research at their own discretion without worrying about time constraints. Thus, we are able to provide certain in-demand library services at all times.

One of the demands that some students have is the ability to work with others to solve a problem or learn a skill. Great Valley students can collaboratively use the online catalog tutorial to understand searching concepts or can consult the library tour when they are discussing a library policy with a study group. The flexibility of the technology used to create these lessons give students the freedom to choose to share learning or learn alone. Technology-based education allows adult students to proceed in learning on their own if they wish. Some adult students who have had little contact with computer technology feel anxious when confronted with an online catalog. The LIAS tutorial allows students to work on tasks anonymously in the privacy of their home or office until they feel more comfortable with searching. This aspect of the technology may be used initially, before the student enters the library and searches for materials in a public area, or it can be used after group library instruction to enhance skills. Technology-based instruction allows the student to take risks that they may not be comfortable taking in a classroom.

Feedback on Great Valley's Internet-based library instruction modules has been positive. Students like the flexibility they have to access it anywhere and to use only the parts of it they need. A few students have commented that they also want the human touch in library instruction but that the library tour and tutorial have still been valuable. Overall comments indicate that it is truly meeting the needs of adult students.

Conclusion

Technology-based instruction can be used to effectively meet the requirements of adult students. It can allow students to learn at their own pace and to choose what they need to learn, including skipping sections that are inappropriate or not needed for their particular requirements. Using the Internet, students can access learning at anytime from anywhere as long as they can connect with the Net. An Internet-based online catalog tutorial like the one developed at Penn State Great Valley can also permit students to immediately use what they have learned by switching to the catalog for actual searches. As a group or individual, depending on the needs/preferences of the student, all of this can be accomplished. When one reflects on the positive aspects of the interface of technology and adult education principles, it quickly becomes evident that more institutions should consider the development of technology-based instruction to meet the needs of adult students.

References

Brookfield, S. (1986). *Understanding and facilitating adult learning*. San Francisco: Jossey-Bass Publishers.

Drexel University College of Information Studies. Alfred P. Sloan Center for Asynchronous Learning. Retrieved July 28, 1998 from the World Wide Web: http://aln.cis.drexel.edu/.

Fischetti, M., Anderson, J., Watrous, M., Tanz, J., and Gwynne, P. (1998, March/April). *University Business*, 48-49.

Knowles, M. (1990). *The adult learner: A neglected species*. Houston: Gulf Publishing.

Lawler, P. (1991). *The keys to adult learning: Theory and practical strategies*. Philadelphia: Research for Better Schools.

Merriam, S. (1991). *Learning in adulthood.* San Francisco: Jossey-Bass Publishers.

Pennsylvania State University. Penn State Great Valley LIAS Search Tutorial. Retrieved July 28, 1998 from the World Wide Web: http://www.gv.psu.edu/library/lias/liasp_1.htm.

Pennsylvania State University. Penn State Great Valley Library Tour. Retrieved July 28, 1998 from the World Wide Web: http://www.gv.psu.edu/library/tour/.

Pennsylvania State University. Penn State's World Campus. Retrieved July 28, 1998 from the World Wide Web: http://www.worldcampus.psu.edu/.

Tice, E. T. (1997). Educating adults: A matter of balance. *Adult Education.* Vol. 9, pp. 18-21.

Zemke, R. and Zemke, S. (1991). Thirty things we know for sure about adult learning. In Jones, P. (ed.) *Adult learning in your classroom.* Minneapolis: Lakewood Books.

CHAPTER 18

Electronic Lawyering and the Academy

William R. Slomanson
Professor of Law

Thomas Jefferson School of Law
San Diego, CA

Reprinted from the Journal of Legal Education © *Association of American Law Schools (AALS)*

This chapter is offered to enrich the dialog about preparing law students for the technical realities of law practice. Some law teachers contend that we have been sluggish about responding to technological needs: "Law schools have been much slower than other professional and graduate schools to adopt computer-augmented teaching methods, perhaps because little evidence has ever been presented to law teachers that the necessary expense and effort can be justified by an improvement in student learning."[1] Until recently I assumed that the technical aspects of the teaching profession were someone else's responsibility. The hallowed halls of academia sheltered me, although the students I served were about to enter a world of client e-mail, law firm Web sites, and electronic filings.[2]

But my assumptions were challenged. I now believe that integrating teaching and technology is my responsibility and may be yours as well. If we can teach students to think like lawyers, we can help students to perform like lawyers. That means helping them make the transition into today's professional world, which already depends on tomorrow's technology.[3]

WHY ME?

At an AALS (Association of American Law Schools) annual meeting, Bob Berring of Boalt Hall advised his audience that "this" would one day be a relic of the past. "This" was the casebook he clutched in his hand, waving it to make his point.[4] I noted that the book he was waving was the very casebook that I have used for nearly two decades. My reaction was predictable: What he was saying had no application to a seasoned legal educator like me. After all, you can be a good teacher whether or not you employ technology in (or out) of the classroom—right? I was not ready to accept the prospect that the attractive volumes we carry to class each day might become tomorrow's dust collectors.

I experienced a remarkable attitude adjustment. I began to question whether I had spent my professional time wisely. I had published articles and books, earned my share of teaching awards, done my share of service to my school (lucky me-tenure chair for these last three years), and crusaded for practice-oriented curriculum changes in the aftermath of the MacCrate Report.[5] I could smugly pat myself on the back at mid-career. But I had a nagging sense that I was overlooking something, and that I had much more to do if I intended to prepare my students for contemporary law practice. Seven days after the AALS conference, I established my first electronic classroom extension-an e-mail discussion group. Seven months later, I constructed my own academic Web site to launch a paperless law class. Only then did I appreciate the role that technology could play in augmenting the traditional classroom experience.

Another beam of insight radiated from that AALS Techie Day: I might one day transcend the limitations of the print world if I seriously explored the seemingly boundless possibilities of electronic education and publication. I thought about how I might incorporate the World Wide Web into my classroom teaching (as well as future publications), so that my classroom materials might be as current as the date. Several innovators have lauded the virtues of integrating technology and teaching, a conspiracy which I am obviously on record as supporting.[6] However, others have cautioned that technology has its limits and even its dangers. The computer may fail where the book succeeds.[7] Peril accompanies promise when training legal writers in the new electronic era.[8] And the age has, to some degree, alienated attorney and client—a "disease...contracted in law school."[9]

Perhaps most important, one should make careful choices about how to allocate one's time. Being on the cutting edge takes time and effort. A reasonable alternative is to wait until the technology is nicely packaged for the less adept. As Northwestern's Dean David Van Zandt has noted:

> A great deal of time can be spent on technological presentations that
> is probably better spent by our tenure-track faculty on scholarship.

This is not to say that technology is not important. It is very important in the modern world. It is just that we have limited resources and must use them most effectively.[10]

Despite those words of caution, it is my hope that you will at least peruse the following summary of what others are doing to incorporate electronics into their law school teaching.

THE ACADEMY IN CYBERSPACE

Other than attending conferences, the most informative sources for quickly surveying the terrain of educational cyberspace are Bernard Hibbitts's JURIST Web site, which collates the worldwide teaching resources that were previously adrift on the Internet; and the sparse but proliferating law review literature on computers and legal education.

THE WEB SITE NETWORK

In 1996, Bernard Hibbitts of the University of Pittsburgh made two significant contributions to the legal academy. One was a provocative article questioning the continued vitality of print and electronic law reviews.[11] The second was his Web site, JURIST: Law Professors on the Web (now JURIST: The Law Professors' Network).[12] JURIST provides instantaneous access to the educational Web pages of law teachers all over the world. We can now compare notes internationally on using the Web for classroom teaching.

There are a number of useful pages on the Jurist Web site. Those most relevant to this chapter are Course Pages, Resource Pages, and Home Pages. Course Pages has HTML links to various Web sites, alphabetized by general academic topic and course. Although many of these course Web sites are exhaustive, others are less ambitious. Most open with introductory remarks about course content, then direct you to other pages within the particular Web site containing material for that teacher's course. Resource Pages is a useful research tool. It collates Web sites containing online resources from other sites that the teacher wishes to associate with his/her course. Home Pages is an alphabetical listing of everyone who has a Web page on JURIST.

LAW REVIEW LITERATURE

During the first generation of law school computing, we generally sat before our machines to produce scholarship or to retype our scribbled class notes. In this second generation, a growing number of teachers are using the computer for classes that are no longer isolated by physical boundaries or even confined to a classroom. Some are augmenting the traditional classroom experience with an e-mail discussion group. Some schools have laptop sections (or entire entering classes) that integrate the physical classroom with the virtual reality of the Internet.[13] And some classes are available in different law school locations via the distance learning made possible by video technology.[14]

There is an emerging body of literature about who is doing what to promote lawyering; which can help us learn more about using technology in the educational environment.[15] We can better serve our clients (students and the public at large) through teaching and library development.[16] One does not have to be a computer guru to incorporate the World Wide Web into legal education. An informative article by Michael A. Geist traces the evolution of electronic legal education from computer-assisted legal education to the Internet. Geist confirms that "the Internet enables legal educators with little or no computer training to experiment with innovative teaching methodologies and, in the process, to combine the best of CALR (computer-assisted legal research), CAI (computer-aided instruction), and electronic casebooks and to excite law students uninspired by traditional law teaching techniques."[17] The dubious wager of another era—would the result be worth the effort?—is no longer a long shot. You can enrich your courses with technology if you are willing to invest some research and development time.

THE NEW FRONTIER

My students prepare motions in state civil procedure. This began as an optional third unit in my California civil procedure class. Ultimately, it may increase in unit value and be offered as a completely unattached course. For the moment, I am somewhat fearful that the lack of any on-campus attendance requirement might raise accreditation concerns. While no one that I know of has posed the issue, it seems prudent to require students to take the on-campus (two-unit) course along with the optional (one-unit) cyberspace course.

Students access the essentials from my Web site. It contains a client interview and some pleadings. They then submit three distinct motions during the semester, which I grade. No paper copy is permitted. All communications are by e-mail (submissions preferably by attachment). Students may ask no questions in person, because I characterize such potential inquiries as ex parte communications with the judge. They may post to the group, to each other, or to me if they feel that clarification is needed—which I make and then post to the entire class. Put another way, they have to do their own work, as in practice. I am available; however, to help with the technical or procedural problems that may arise.

My students, especially those who travel some distance to get to school, are ecstatic for two reasons. I am available at all hours of every day, and they can do virtually everything they have to do—including submitting their bluebook in their bathrobes. As this new frontier evolves, you may need to explain patiently to your dean that you can provide valuable service while you are not on campus. If access to the instructor is of value to students, electronic availability sometimes makes more sense than being in your office (perhaps with a horde of students at your door competing for attention). Many a student/teacher conference can be conducted by e-mail, without disrupting either participant's schedule.

E-MAIL COMPONENT

Among the first questions to consider, regardless of your subject area, is whether or not to establish an e-mail discussion group for your course. You might at least conduct a bulletin board style, question-and-answer forum. That is a minimal investment with a big return: all students benefit, both active participants and parasitic lurkers. You don't need technological support to do this; anyone who has sent only one message can set up an e-mail discussion group for every section of every class without any assistance from a commercial vendor.

A more ambitious step, like incorporating the Internet into your class(es), may necessitate a weighing process. Will your institution look kindly on your venture into the e-world, or will you be frowned upon for turning your attention away from scholarship, service, and the other traditional time commitments? Obviously, any non-tenured teacher must consider such questions, and even a senior tenured professor should probably not ignore them. But adding an e-mail component to your course is no radical venture. It would be a useful starting point for the fainthearted.

An e-mail discussion group can advance pedagogical objectives before, during, and after class.[18] Before class, you can post questions to provide some basic guidance about the key issues that you will cover. During class, you and your students will be on the same page from the outset. After class, you can be available via e-mail for electronic conferences. Thanks to e-mail, you can continue interesting discussions for which there was insufficient time in class, post hypos for student discussion or review, and answer the typically short questions that students may raise after you have left the classroom.

This method of extending the confines of the physical classroom received a great deal of attention because of the West's fall 1997 release of its TWEN (The West Educational Network.) project, which allows a teacher to conduct e-mail discussions and place Web content on a page provided for $20 per student. Message threading in your school's e-mail system may not be as slick as it is in TWEN.[19] So if you feel technologically challenged, you should consult your school's TWEN representative, or whomever is tasked with providing TWEN training at your school.

The Lexis alternative to TWEN is WCB (Web Course in a Box). Unlike TWEN, this system has no costs, and no contracts requiring the law school to indemnify the vendor for actionable student messages. In creating WCB, Lexis drew from a pre-existing university Web template, modifying it to provide similar e-mail and Web posting services for law teachers.[20]

There may be downsides to TWEN or WCB.[21] Some of us may be concerned about inducing student reliance on a particular commercial vendor. One can accomplish roughly the same objective without TWEN or WCB, with little technical expertise beyond what it takes to send an e-mail message. The Address Book function of the garden variety e-mail system conveniently permits teachers (and students) to create and participate in many different e-lists, not just the one established by the teacher. You can dispatch messages to all list members with only one address in the Mail To line of the e-mail message. The teacher who wishes to generate a message to even the largest of classes can send bcc copies to all list members. My students and I address discussion group e-mails to LIST, with the name of the particular class (e.g., FEDCIVPRO, CALCIVPRO, or INTLAW) at the beginning of the Subject line. Students who see from the Subject line that this message is of no interest to them need only hit the delete key. Careful instructions about the Subject line can accomplish a degree of threading that is less glitzy than TWEN's or WCB's, but perfectly useful.

Regardless of which e-mail system you and your students have already selected, you should personally survey the extraordinary e-mail programs, easy to use and install, that simultaneously operate with any e-mail system from an Internet service provider (ISP) much as the Windows enhancement operates with DOS. If you would not return to DOS-based computing, after operating in a Windows environment, you should evaluate one of these e-mail enhancement programs. They are more attractive and more functional than the generic low-grade e-mail system. You can access multiple windows at once, automate routine tasks, and improve the likelihood of successfully sending and receiving attachments.

WEB SITE CONSTRUCTION OPTIONS

There are four choices for those who wish to construct an academic Web site:

- the .edu option—the simplest alternative, because you merely add course content to your individual faculty page on your school's existing Web site

- the TWEN/WCB option described above—to add textual content to your individualized West TWEN or Lexis WCB page

- the .com option—large ISPs offer free Web sites as a strategy for encouraging more costly upgrades[22]

- the do-it-yourself option—arguably the most complex Web site construction option, with the compensating advantages of not having to rely on someone else's priorities to get new material posted promptly (.edu option) or inducing student reliance on a commercial vendor (TWEN/WCB options)

These options are more fully analyzed in the initial "Lessons from the Web" feature on JURIST: The Law Professors' Network.[23]

COURSE CONTENT

Planning a Web component for your course requires the early resolution of some pedagogical issues. In this section I raise what I consider critical questions, with suggested answers and alternatives.

Should you get into e-teaching, just to add a technological component to your course? Although my own attitude toward teaching technology is effervescent, I cannot naively anticipate that all readers of this chapter will be equally enthusiastic or will bypass technology for the same reasons. A sense of balance compels me to offer the following advice: First, Do No Harm! One must be exceedingly cautious about jumping onto a technological bandwagon with no clear sense of direction. The teacher should carefully weigh the course content to be sure that any infusion of technology will really serve a useful purpose.[24]

What is your primary pedagogical objective in bringing technology into your course? There are many alternatives and no clear favorite. You may wish to do one or more of the following: augment your classroom experience (with an email discussion group); Webify the in-class experience (with Web presentations or laptop-based course material); or avoid the physical confines of the traditional law school class (with distance learning or a paperless e-class).

For example, the main objective in my e-class is to make the students more proficient in state motion practice via technology they will use as lawyers.[25] I want to improve their general technical know-how so that they can send me an e-mail attachment containing the motions they will encounter early in practice. It is only a matter of time before all pleadings and motions will be submitted to the courts and exchanged between the parties via e-mail attachment.

What are your related objectives for incorporating technology into your course? My complementary objectives evolved on two levels. On the surface, I wanted students to take another skills course while in law school and leave with unique writing samples for their résumés. At a deeper level, however, I envisioned practical exercises in solving problems—both problems of substance and problems of technological mechanics. I wanted students to apply reasoning in a skills course context and I hoped for more interaction with my students than is possible in the classroom. Regardless of your subject, you can formulate a variety of supporting objectives to fit any course or pedagogical approach.

What course content should you put on your Web site? This option separates an e-mail discussion group from an e-class. The e-class page on my Web site contains a short factual scenario in which law students contract food poisoning from eating Edible Widgets at a deli near the university. This Web page links to a separate Web page containing the hypothetical that spawns all semester assignments. Students then click on Problem One, Two, or Three, which whisks them to the various problems/motions that they will prepare. The students submit their graded motions to me via e-mail in the fifth, tenth, and fourteenth weeks of a fifteen-week semester.[26]

A word of advice about what you ask your students to do electronically: Never require them to do anything that you have not already done yourself. For example, I made the submission of motions on 28-line pleading paper a requirement for my first e-class. I quickly learned that this should have been optional. I knew that it could be done, but I didn't know how. When I was asked, I had to confess ignorance. At that point a technologically-gifted student figured out how to do it and e-mailed detailed instructions to me and to the rest of the class. Should you prefer to save your confessions for your place of worship, then privately seek advice from students or staffers before you require task X to be performed. You can then put how-to-do-it advice on your Web site (e.g., by creating a separate Web page for frequently asked questions).

What administrative information should you include on your Web site? Only one's experience and imagination limit this option. For my e-class, I set forth the honor code requirements, the grading system, withdrawal options, a description of the instructor's role, various e-mail addresses for submitting questions about the course, and my netiquette requirements. For my other courses, I constructed a course page where students can find an array of information. From each course page, they can link to current reading assignments, prior exams, assigned problems not in the casebook, and other Web sites providing useful resources in that field. From my international law course page, for example, they can link to my A to Z listing of international Web sites for students and lawyers. From another course's page, they can link to the Web page where I post new cases previously posted to interested graduates.

Should you grade on a non-anonymous basis and/or employ a staff assistant? Anonymity is best, of course, but it can present a host of problems in an electronic environment. Because my course is paperless, it will remain "experimental" for some time. There will be a batch of new recruits in every e-class for the next few years, with varying levels of technological sophistication. I will have to be flexible in my expectations. Undergraduate experience provides ample computer knowledge for general purposes, but it doesn't give students the specific tools they need for legal education.

More specifically, students must have an ISP to register for my course. Then, they must get a second e-mail account (student identity unknown to the instructor). They use the second account for submitting their assigned problems.[27] This system allows anonymous grading, but it's not foolproof. Murphy's Law also applies to ISPs.[28] The second e-mail account may not be available to the student at the last moment before the deadline for submission. Students then have the option of submitting the assignments via their known e-mail addresses, which we use for class administration and communication during the semester.

One might wonder whether or not students could submit their assignments to a third party (a staff assistant), who would strip their e-mail addresses before sending the assignments to the instructor. That might seem simpler than requiring a second e-mail account, but submitting assignments to someone other than the instructor totes several risks. One is that you cannot really know whether or not the paper was submitted on time. I require (what appear to be) inflexible deadlines to maintain an aura of reality approximating actual law practice. But many e-mail systems do not give you real-time posts based on your time zone. For my very first e-class motion, which was due at midnight on the fifth Saturday of the semester, I checked mail at that precise moment. I had received one submission that showed a sending time of 5 a.m. the following day. Because there are few countries with five time zones, I suddenly recognized the dangers of asking another person to monitor the mailbox. One could always choose a submission date and time which coincides with the supposed presence of the staff assistant, but what are the odds that the assistant will monitor submissions as closely as the instructor (particularly if that staff person must juggle many priorities, not just yours)?

A staff member does assist me in a more limited way. She links the students' anonymous e-mail addresses with their real identities after they have submitted a motion and I have graded it. Thus, she has the difficulty of dealing with those students who fail to follow requirements like using the same e-mail address to submit all of their course problems. That recurring nightmare produces more e-mail blue-book addresses than the number of students registered.

Should you archive your e-messages and/or e-class submissions? Your first thought may be that retaining an archive will dampen students' willingness to post to the list: The only thing worse than posting e-mail for all to see is knowing that all will see it for time immemorial—or however long the instructor chooses to archive messages. One response to this concern is that archiving makes a student carefully consider sending rude or vulgar messages. Archiving also allows the procrastinator to catch up; he/she can access all the e-mails posted during the first few weeks while he/she was course shopping.

At this stage in the evolution of my paperless e-class, I have not posted student answers on the course Web page. There are two reasons. First, my students appear to be a bit nervous about worldwide access to their ungraded motions or to the version that includes my sometimes unflattering commentaries. Second, I prefer to operate my experiment a little while longer before I consider compromising the work product that I generate. Put another way, when I have no more revisions to make to my e-class problems, then that might be the time to place student answers on the Web.

Another reason for archiving is to create a safety net. I administer my e-class from home. My students submit their messages and motions to both my home address and my office address. On the day that the students' first motion was due in my first e-class, my home system suddenly crashed (the cyberspace version of Murphy's Law). I had received only about a third of the electronic bluebooks when this occurred, but I was able to go to school and retrieve the archived copy of the remaining bluebooks. Without an archived copy, I could not have determined whether or not those other students had filed their motions by the deadline. And I would have had to delay grading until they resubmitted their motions—not exactly state-of-the-art procedure.

The final advantage to archiving is that I can conveniently forward my archived bluebooks for institutional storage to the staff member who retains copies of all print—and now, electronic—bluebooks. Perhaps one day all bluebooks will be electronic and we will no longer have the hassle of storing and retrieving reams of paper.

Can students take the class if they aren't wired? You should consider having a computer available in the library, or elsewhere in your building, for sending and receiving e-mail messages and class assignments. In effect, I have two on-campus archives: one in my office and one in the library. Students may use the library's computers to send and receive messages (and anonymous bluebooks). That serves two purposes. One, it allows access to my course for those vanishing diehards who don't own a computer. Two, students can use the library computers when they are facing a deadline and their e-mail system suddenly fails.[29]

What Web pages might you create, even if you don't launch an e-mail discussion group or a paperless e-class? When I began to consider using the Web for teaching, I constantly thought of new uses for my Web site. For example, I created a page listing my publications. This has saved me a lot of paper and time when I'm asked to provide that list.

Essentially, I began by constructing teaching-related pages: Purpose, TechInfo, FAQ, Courses, Reading, Exams, Problems, and NewCase. These pages serve assorted teaching objectives. The Purpose page gives a short explanation of each page on my Web site; it is a convenient navigational tool that helps readers decide which pages are relevant for their needs. The TechInfo and FAQ pages help the students in my e-class overcome common technological hurdles. The Courses, Reading, Exams, and Problems pages all help me avoid the drudgery of providing the same information to students over, and over, and over again. NewCase is where I place new cases of value to my various civil procedure courses. It is also a convenient place to gather candidates for subsequent editions of my classroom texts in California civil procedure and international law.

Should you install a security system that requires a password from anyone seeking to access your Web site? If you have chosen either your school's .edu Web site or the do-it-yourself Web site that I recommend, you might consider whether you need a so-called firewall to keep out students, or others, who might inject flames or viruses. I don't intend to do this until delusions of grandeur suggest that my technological universe is a bit larger than its current magnitude.

Can you use your Web site for publication purposes too? There are at least three arguments against e-casebooks and e-articles. One is the obsession with leather bound work products to display on the mantle, not to mention the stunning university seal on law review reprints. Another is the institutional resistance to self-publication, which lacks the imprimatur of a recognized commercial vendor or law review. Finally, there is that ego-driven insistence on keeping track of citations to one's work. Of course having a Web counter on your Web site, or asking your research assistant to look up Web citations to your work, will yield results similar to those you routinely get through traditional channels.

As we continue to explore this new technology, more of us will discover the value of an e-book or online article as opposed to traditional print material.[30] Obsolescence will itself become obsolete. You will make changes instantaneously, rather than waiting for the next edition or the annual supplement. You will also be able to link to legal resources elsewhere on the Internet. This can bring your course book to life in ways never dreamed of in the traditional world of print publishing.

CONCLUSION

A Web-based class will be, initially, more work for the teacher than a traditional class with comparable unit value. This conclusion may be verified when you suddenly find yourself treading water with those students who are less apt to take to the technological waters than the others. The offsetting advantage is that you can benefit both the student and your career development. You can first augment a traditional law school class with an e-mail discussion group; later devise a paperless e-class to integrate e-mail, the Web, or other computer technology; and ultimately explore the advantages of publishing the first (or next) edition of your own textbook on CDROM or your Web site.

If you believe that we have a professional obligation to prepare our graduates for the practice of law, then the innovative opportunities offered by technology are irresistible. Having been schooled in a generation of passive learning, I had previously experimented with the usual ways of altering that classroom dynamic: I tried in-class moots, for example, and I incorporated the problem method into my courses. But nothing has presented more satisfying results than my Web- based teaching experiment.

I will conclude by identifying two core values that are my rationale for taking this stimulating plunge. First, a class with an electronic component offers a new beginning. It presents exciting interactive opportunities for supplementing the traditional classroom experience. It can provide a refreshing perspective on the virtually unlimited resources available on the Internet. I am not sure that I agree with my distance-learning colleagues who predict the demise of the classroom as we know it; I still see technology as playing a supporting role, not signaling a casebook/classroom apocalypse. But time will tell whether or not Bob Berring's predictions will come true.

Second, my students and I have found that our Web course experience is very near to the real thing. My students work from their home or office computers, just like real lawyers, without having to go to court/class every day. They engage in collaborative learning via private and group e-mail messages about their pending assignments, just like real lawyers in real practice. They have unyielding deadlines, just as they will have in what they so fondly refer to as "the real world." One may, of course, accomplish similar objectives in a more traditional classroom environment. But all my students emphatically assert that this is as real as it gets in law school without actually going to court.

At this point I could provide self-serving excerpts from students' course evaluations. Instead, I have asked some of my students to serve as resources if you have specific questions about their reaction to whatever scheme you might devise. They too are pioneers, and I thank them for contributing to the success of my paperless e-class experiment.[31]

William R. Slomanson thanks the following for their comments: Glen Peter Ahlers (Arkansas), Michael Geist (Columbia), Marybeth Herald (Thomas Jefferson), Bernard Hibbitts (Pittsburgh), Andrea Johnson (California Western), John Rogers (Kentucky), Stephen Sowle (Chicago Kent), and David Van Zandt (Northwestern). Special thanks to Kay Henley, director of academic administration at Thomas Jefferson School of Law, for her assistance during the first e-class.

References

1. Paul F. Teich, How Effective Is Computer-Assisted Instruction? An Evaluation for Legal Educators, 41 *J. Legal Educ.* 489, (1991).

2. In December 1997 Microsoft (and three other companies) announced a plan to develop an electronic filing system for all United States and Canadian courts. This project will automate the entire legal system-from each law firm to every judge's bench. Anthony Aarons, Microsoft Sets Its Sights on Electronic Filing Systems, Cyber Esq. Supplement to legal newspapers including the *Los Angeles Daily Journal* and the *San Franciso Daily Journal*, winter 1998.

3. See generally Ruta K. Stropus, Mend It, Bend It, and Extend It: The Fate of Traditional Law School Methodology in the 21st Century, *Loyola University of Chicago Law Journal*, 27, 449 (1996).

4. The Big Picture (remarks ending workshop on Teaching and Technology in the Twenty-first Century (Washington, January 1997). As Berring noted several years earlier: "It is amazing that in 1993 law students still read casebooks, desiccated collections of appellate opinions. But rather than being the introduction to the old world of the case categorizations of West, the textbooks now are the last bastion of the case gestalt. The rest of the research world is changing." Robert C. Berring, Collapse of the Structure of the Legal Research Universe: The Imperative of Digital Information, 69 *Wash. L. Rev.* 9, 32 (1994). A telling history is available in Steve Sheppard, Casebooks, Commentaries, and Curmudgeons:An Introductory History of Law in the Lecture Hall, 82 *Iowa L. Rev.* 547 (1997).

5. Section of Legal Education and Admissions to the Bar, American Bar Association, Report of the Task Force on Law Schools and the Profession: Narrowing the Gap, Legal Education and Professional Development—An Educational Continuum (Chicago, 1992). My crusade is

reported in Richard C. Reuben, A Man with a Mission: California Law Prof, Touts State Civil Procedure Courses, *American Bar Association Journal*, (Feb. 1997) p. 34.

6. Richard Warner, Teaching Electronically: The Chicago-Kent Experiment, 20 *Seattle U. L. Rev.* 383 (1994); Scott A. Taylor, Teaching a Law Seminar over the Internet, 7, *Journal of Law and Information Science*, p. 41 (1996).

7. Richard Haigh, What Shall I Wear to the Computer Revolution? Some Thoughts on Electronic Researching in Law, 89 *Law Library Journal*, p. 245 (1997).

8. Lucia Ann Silecchia, Of Painters, Sculptors, Quill Pens, and Microchips: Teaching Legal Writers in the Electronic Age, 75 *Nebraska Law Review*, p. 802 (1996).

9. Robert H. Thomas, "Hey, Did You Get My E-Mail?" Reflections of a RetroGrouch in the Computer Age of Legal Education, 44 *Journal of Legal Education*, p. 233 (1994).

10. E-mail to author (February 6, 1998). I construed his 1998 AALS presentation (as part of a panel on Funding Legal Information Technology in the Law Schools) as the most compelling acknowledgment that the teacher and the institution must base the desire to integrate teaching and technology on something more than the glitzy factor of being in the technological vanguard.

11. Last Writes? Reassessing the Law Review in the Age of Cyberspace, 71 *New York University Law Review*, p. 615 (1996) (also at <http://www.law.pitt.edu/hibbitts/lastrev.htm>). The responses to Hibbitts's predictions include Henry H. Perritt, Jr., Reassessing Professor Hibbitts's Requiem for Law Reviews, 30 *Akron Law Review*, p. 255 (1996). For a reply, see Bernard Hibbitts, Yesterday Once More: Skeptics, Scribes and the Demise of Law Reviews, 30 *Akron Law Review*, p. 267 (1996) (also at <http://www.law.pitt.edu/hibbitts/akron.htm>).

12. See <http://jurist.law.pitt.edu/index.htm> (revised version, February 1998). Both Cambridge University and the Australian National University operate JURIST "mirror" (affiliate) sites, meaning that JURIST is no longer just a United States Web site. It is a global network of law teachers.

13. For a listing of laptop schools, see James E. Duggan, Mandating Computer Ownership at Law Schools: A Survey, at <http://www.siu/offices/lawlib/survey.htm>.

14. See, e.g., Andrea L. Johnson, Distance Learning and Technology in Legal Education: A 21st Century Experiment, 7 *Albany Law Journal of Science and Technology*, p. 213 (1997); Susan E. Davis, Remote Learning by Leaps and Tumbles, *California Lawyer*, (August 1996) p. 49. On the benefits and burdens of distance learning, see Jill M. Galusha, Barriers to Learning in Distance Education, 5 *Interpersonal Computing & Technology Magazine*, (December 1997) p. 1. (Articles stored as files at Georgetown University). To retrieve this file (article), place the term GET both before and after <GALUSHA.IPCTV5N4> in the body of your e-mail message to <LISTSERV@LISTSERV.GEORGETOWN.EDU>.

The A.B.A. Distance Learning Guidelines are contained in a letter from James P. White, consultant on legal education to the American Bar Association to deans of ABA Approved Law Schools (May 6, 1997). Fortunately, they do not purport to apply to an e-class at one law school, although more miles than the distance may separate teachers and students between the two schools involved in a distance-learning project.

15. See, e.g., Rosemary Shiels, Technology Update: Attorneys' Use of Computers in the Nation's 500 Largest Law Firms, 46 *Am. U. L. Rev.* 537 (1996); Barbara Bintliff, From Creativity to Computerese: Thinking Like a Lawyer in the Computer Age, 88 *Law Libr. J.* 338 (1996); Ethan Katsh, Digital Lawyers: Orienting the Legal Profession to Cyberspace, 55 *U. Pitt. L.*

Rev. 1141 (1994); William T. Braithwaite, How Is Technology Affecting the Practice and Profession of Law? 22 *Tex. Tech L. Rev.* 1113 (1991); Twenty-ninth Selected Bibliography on Computers, Technology and the Law, 23 *Rutgers Computer & Technology Law Journal*, p. 419 (1997).

16. See Eugene Volokh, Computer Media for the Legal Profession, 94 *Mich. L. Rev.* 2058 (1996); Richard A. Matasar & Rosemary Shiels, Electronic Law Students: Repercussions on Legal Education, 29 *Val. U. L. Rev.* 909 (1995); Richard A. Danner, Facing the Millennium: Law Schools, Law Librarians, and Information Technology, 46 *Journal of Legal Education* 43 (1996); Gail M. Daly, Law Library Evaluation Standards: How Will We Evaluate the Virtual Library? 45 *Journal of Legal Education*, p. 61 (1995).

17. Where Can You Go Today? The Computerization of Legal Education from Workbooks to the Web, 11 *Harvard Journal of Law and Technologies*, p. 141 (1997) (also at <http://jolt.law.harvard.edu/articles/11hjolt141.html>).

18. Stephan D. Sowle & Richard Warner summarize the advantages:

[Use of e-mail discussion groups] frees up additional classroom time for delving deeply into theoretical and policy considerations, exploring hypothetical applications of the legal rules under discussion . . .Classroom time can, thus, be reserved for discussion of difficult or sophisticated issues that are best taught in direct exchanges with students.

Another benefit is that students who lack the confidence to participate in class may feel more comfortable taking part in online discussions, particularly if they can be anonymous. This helps counteract the unfortunate tendency for classroom discussion to be dominated by a few students.

Cruising the Electronic Classroom, *Law Teacher*, (fall 1997) p. 4, 5.

19. Threading refers to an e-mail system's capability to arrange (and rearrange) incoming e-mail messages by a particular order (date, sender, topic, etc.).

20. One might surmise that, because the comparable Lexis WCB is a no-cost alternative, West will ultimately drop its plan to charge for TWEN—or abandon TWEN altogether, as more and more people choose to construct their own Web sites.

21. You may want to tell your students that the use of TWEN or WCB inserts a "cookie" into their computer programming, making it possible to gather information about the computer user's tastes and usage of the product.

22. I upgraded to the unlimited Internet access pricing option with my ISP, merely to obtain a no-cost host for the academic Web site that I was planning. I do not recommend the "cookie cutter" templates offered by ISPs, such as AT&T and AOL. These are fine for many purposes, but not for a state-of-the-art law course.

23. See William Slomanson, The Four Corners of the Academic Web Site World, at <http://jurist.law.pitt.edu/lesfeb98.htm>. Note: Do not dismiss the do-it-yourself option because you think that do-it-yourselfers have to understand the intricacies of HTML coding. There are marvelous Web site construction programs that eliminate the drudgery associated with inputting HTML coding. The most popular program is probably Adobe Page Mill. See <http://www.adobe.com/prodindex/pagemill/main.html>. Using TWEN or WCB is easier, but the do-it-myself option has given me a much greater return on my time investment.

24. I don't recommend requiring students to take part in an e-mail discussion group or grading them on their participation.

25. They prepare state practice motions because the vast majority of students will practice in state courts long before they ever appear in federal court—or before they decide to pursue a predominantly federal practice.

26. See e-class Web page at <http://home.att.net/~slomansonb/eClass.html>.

27. There are several companies that offer free e-mail service so that students do not have to pay for a second e-mail account just to take my course.

28. During the semester when I operated my first electronic discussion group, AOL was down for a long period. (AOL had advertised an attractive upgrade, without anticipating the new demands on its existing system as millions of people joined.) Many of my students learned the hard way that ISPs are not always up and running.

29. A number of our colleagues will dispute the propriety of allowing anonymous e-mail messages from a non-traceable source. Because of actual, perceived, or potential student abuse-the subject of past LAWPROF or ETEACH threads-some teachers do not allow anonymous postings, or any submission that the teacher has not screened.

30. See Gregory E. Maggs, Self-Publication on the Internet and the Future of Law Reviews, 30 *Akron Law Review*, p. 237 (1996).

31. Their personal e-mail addresses are: cgorian@juno.com; hangellea@hotmail.com; hsbass@juno.com; jd@thegroup.net; lido14@worldnet.att.net; mischmart@aol.com; tpmoye@aol.com; vancho@inetworld.net (Professor: bills@tjsl.edu).

CHAPTER 19

Study Abroad for Science and Engineering Students via Distance Learning

Leonard A. Van Gulick
Professor of Mechanical Engineering

Lafayette College
Easton, PA

Jack R. Kayser
Educational Technologies Consultant

Nazareth, PA

INTRODUCTION

The distance-learning program was initiated at Lafayette College in 1996 with grant support from the Andrew W. Mellon Foundation and the Center for Agile Pennsylvania Education (CAPE). Initially the program was a cooperative arrangement among several private colleges in the Lehigh Valley of Pennsylvania to develop and share classes via ISDN (Integrated Services Digital Network) line-based teleconferencing. As the program grew, it incorporated connections to institutions outside Pennsylvania.

A one-semester Lafayette study-abroad program for engineering students has been in place at the Free University of Brussels (VUB) in Belgium since 1990 (Van Gulick and Paolino 1997). Students can participate in the program and still complete the requirements for any Lafayette engineering degree in the usual four years. This requires that students in Brussels be able to take the engineering courses necessary to avoid falling out of sequence in the highly structured engineering degree programs.

The Lafayette distance-learning program has been extended to include communication with study-abroad students in Brussels. The first engineering course, Strength of Materials, was transmitted from Lafayette to the VUB via ISDN lines in the spring of 1997. The course was again transmitted in the spring of 1998, and is now a standard component of the study-abroad program. The use of video conference technology in the course is part of a larger effort on the part of the authors to incorporate electronic communications in the Lafayette engineering curriculum [Kayser 1996].

DESCRIPTION OF DISTANCE-LEARNING COURSE

The course will be examined from two perspectives: teaching and technology. From the teaching perspective, the course occurs over a 14-week semester. The instructor has three hours of ISDN-based teleconference contact with the entire class each week. One and one-half hours are spent on Tuesdays lecturing and answering student questions, while on Thursdays, one-and-one-half hours are devoted primarily to interactive discussion. Course lectures are structured around class notes, which the instructor displays using a document camera. Frequent camera changes, from the document camera to a camera showing the instructor, maintain interactive contact with the students.

Class notes are placed on the Web in a password-protected format at least several days prior to each class. Paper copies of the notes are printed in Brussels and provided to the students. Students bring the notes to class and add annotations and comments to them during the lecture. Having the notes available frees the students from having to transcribe figures and equations from the monitor. They can, instead, focus on understanding the material being presented and ask questions of the instructor when they have difficulty doing so. This mode of classroom learning has proven to be more time-efficient, as well as more effective overall, than the traditional passive lecture and note-taking mode of learning.

Although assigned homework is an integral part of the course, it is not collected in the usual sense. Instead, the students use the Thursday discussion period to present their solutions to assigned homework problems to the instructor and their classmates using a document camera. This process requires that they complete each assignment on time to avoid publicly disappointing the instructor and their classmates. It also requires that they thoroughly understand their work and document it with clear, legible diagrams, equations, and computations. Requiring the students to present their homework solutions in class has proven to be an excellent way of encouraging a higher standard of performance than in normal on-campus courses. It has also influenced the manner in which the authors teach non-distance-learning courses.

Office hours are held on Wednesdays for students who have questions. These sessions are particularly useful in helping students to overcome difficulties they have encountered in completing their assigned homework. The sessions are conducted using desktop Internet-based video conferencing. The students use a document camera to show the instructor their in-progress homework solutions. The instructor suggests changes and, as appropriate, shows the student portions of the correct solution to the problem under discussion. As during classroom sessions, frequent changes from document cameras to cameras showing the instructor and the students help to maintain close contact during the session.

Further student-faculty interaction also takes place via e-mail throughout the week. Often a short response to an e-mail question is all that is necessary to get a student "back on track" in solving a homework problem. Homework solutions are provided to the students via the Web after the in-class presentations. As with class notes, the homework solutions are password protected. Exam solutions are similarly provided.

The textbook (Hibbeler 1997) played an important role in the 1997 and 1998 offerings of the course. The students purchased their books before leaving the United States and used them extensively, both in and out of class. The instructor frequently annotated the text and displayed it in class using the document camera. Students were expected to simultaneously refer to their personal copies. They frequently added similar annotations. All homework problems and readings were assigned from the text.

The course can only be successfully taught because of the supporting technological resources. There is a direct link between teaching the course and several important elements of technology. The most important technological element is the ISDN-based classroom videoconference system. The system used to date is a VTEL Media Max, running at a 128K transmission rate. The 128K rate proved satisfactory for in-class use, provided no rapid movement occurred when the document camera was in use. This low rate (utilizing two 64K B channels) was used because of the high cost of international ISDN transmission. Even though the class was transmitted at 6 a.m. Eastern time, the telecommunications cost was \$3.88 per minute. This cost is primarily responsible for the authors' efforts to implement complementary Internet-based video conferencing alternatives.

The desktop Internet videoconference system used by the instructor for office hours consisted of an 8500 PowerMac running the Cornell University freeware version of CU-SeeMe. A general-purpose video camera served as a somewhat makeshift document camera. Students at the VUB used a similar desktop system.

Office hours were held between 8 a.m. and 9 a.m. EST. No unacceptable Internet bottlenecks were encountered. The bit rates for transmission and reception were set to their maximum values, and the image was set to the one-quarter screen, low-resolution mode. Normally, only the video and audio functions were used. Occasionally, the chat facility would be used if an audio problem occurred. The whiteboard and document transfer options were not used.

An Internet-connected personal computer was also available in the ISDN videoconference room. This computer was equipped with a basic video camera and CU-SeeMe. It was kept in standby videoconference mode during class in case problems with the ISDN video system were encountered. On one occasion, having this backup capability proved invaluable.

The availability of technical support personnel at both the Lafayette and VUB sites is extremely important. Technical support personnel assisted with video equipment operation and maintenance, class materials preparation and distribution, Web site management, exam proctoring, and scheduling. The class can function only with the committed support of the technical staff, particularly when equipment failures or line drops occur.

Although negligible in the overall cost, the document cameras used with both the ISDN-based and the desktop Internet-based videoconference systems play a vital role. These cameras allow images and calculations to be viewed at both sites, making possible the graphical communication essential in engineering instruction.

The Web site for the class provided the principal means of sending documents to the students (ES230). It served as a file cabinet that could be accessed from both Lafayette and the VUB. Students did not add anything to the content of the site, but instead used it as a source of information. All communication from the students to the instructor took place during videoconference classes, during office hours, or via e-mail. The notes and solutions portion of the Web site was password protected. Only the instructor and the students at the VUB could view the pages. Locking the site was important because the schedule for the students taking the same class on Lafayette's campus was sometimes one or two weeks behind that of the students at the VUB.

Exams were faxed to a faculty member at the VUB who served as an on-site proctor. Completed exams were sent back to the instructor at Lafayette for grading, using a two-day international courier service.

FUTURE PLANS

Teaching two semesters of Strength of Materials to Lafayette students in Brussels has proven that teleconference technology can provide a feasible means of conducting intercontinental engineering classes. Two concerns must, however, be addressed before teleconferencing can be employed on a large scale to expand study-abroad opportunities for engineering students: cost and additional study-abroad locations. Class transmission costs must be reduced to an acceptable level. The two ISDN lines currently used cost nearly $4 per minute of class time. At this rate, expenses may become too high to justify continuing the program in its current form and certainly will not permit its expansion. The possibility of extending teleconference capability to additional locations in Europe or elsewhere must be investigated. This would offer engineering students a much wider choice of study-abroad locations.

The Internet is already being used to reduce transmission cost by employing the CU-SeeMe software for office hours communications. If Internet-based teleconferencing can also be used for classroom communications, transmission costs will cease to be a consideration. Implementing several additional teleconference components will enhance Internet communications. The first enhancement will be the addition of a document-dedicated camera to the CU-SeeMe instructor workstation. A Windows NT computer running the commercial Enhanced CU-SeeMe, which is capable of supporting multiple users, will replace the PowerMac computer itself. This enhanced workstation will be located within the Mechanical Engineering Department and provide the optimum user environment for the instructor.

The student workstation will be upgraded from the current desktop model to a portable laptop computer that integrates a video digitizer, a portable document camera, and a cellular modem. Student users will be given an international Internet Service Provider account. Once properly configured and fully operational, this workstation should be useable anywhere, either within the European community or elsewhere.

The implementation of the enhanced faculty workstation and multiple, portable student workstations should allow several students, perhaps at different locations, to simultaneously participate in the Strength of Materials class using Internet-based communications. With these technological components in place, transmission costs will be drastically reduced and the constraint of a single fixed study-abroad location eliminated.

Obviously, the future plans for this project are predicated on state-of-the-art technologies. These technologies, however, were available at the time this chapter was written.

As improvements occur in processor speeds, system integration, and communication bandwidth, the configuration of the system's technical components will be changed. The most important component in the teaching process, however, will not. That component is the structure and protocol of conducting the class. The format that we have established of interactive discussion, student presentation, and Web content delivery will stay basically unchanged, regardless of changes in the technological details.

References

CAPE, Center for Agile Pennsylvania Education, http://www.acape.org/.

CU-SeeMe, http://cu-seeme.cornell.edu/Welcome.html.

ES230, Strength of Materials, http://www.lafayette.edu/kayserj/es230/index.html.

Hibbeler, R.C., Mechanics of Materials, 3rd ed., Upper Saddle River, NJ: Prentice Hall, 1997.

Kayser, J., "Integration of the World Wide Web and Distance Education in the Engineering Curriculum," Proceedings of the American Society for Engineering Education Mid-Atlantic Conference, ASEE (American Society for Engineering Education) Washington, D.C., 1996.

Van Gulick, L. and Paolino, M., "Internationalization of the Lafayette College Engineering Curriculum," Proceedings of the American Society for Engineering Education Annual Conference and Exposition, ASEE (American Society for Engineering Education) Washington, D.C., 1997, (CD-ROM).

CHAPTER 20

3-D Computer Modeling in a First-Year Architecture Design Studio

Myriam Blais
Professor of Architecture

Université Laval
Québec, Canada

In most architecture schools, design studios constitute the core element of teaching and learning. Design is an activity of environmental transformation by which something is created that was not there before. Since its efficiency relies on successive approximations and iterations, design aims at developing the students' abilities to discover, integrate, apply, and share knowledge. Design is an ongoing process that needs constant practice; architecture students usually enroll in one design studio every semester during the course of their undergraduate studies. Upon entering an architecture school, beginning students, thus, face a dual challenge. They must be simultaneously initiated to design methods, which represents a completely new learning experience for most of them, and become familiar with the appropriate means of presentation. The production and good use of comprehensive, eloquent drawings and models are necessary tools that they need to quickly acquire in order to initiate, reflect upon, develop, and present their design proposition.

While not questioning the relevance of traditional design studios to architecture education, an experiment was conducted in our school during the 1998 winter semester in order to enhance this tradition and make a fuller use of its pedagogical potential by introducing 3-D computer modeling in a first-year design studio.

ARCHITECTURAL DESIGN IN THE STUDIO

In *Building Community—A New Future for Architecture Education and Practice*, Boyer and Mitgang suggest that architecture studios constitute a privileged setting for the refinement of knowledge because they involve a reflective practice that fosters "the learning habits needed for the discovery, integration, application, and sharing of knowledge over a lifetime." Through active inquiry, learning by doing, working within constraints, and collaborative learning, design is meant to generate knowledge and to communicate the "value of architecture to everyone involved and affected by the built environment" (Boyer and Mitgang 1996).

In a similar way, design may be seen as a problem-solving activity. "Students are given a site and a functional program; they transform the first according to their understanding/ interpretation of the second, presenting their propositions by means of drawings and models. There is, however, an unclear side to design because the actions needed to obtain what is expected or desired through such transformation are not "immediately obvious" (Rowe 1992); nor are the solutions to a design problem "strictly true or false." One travels or spirals towards "a domain of acceptable responses" (Zeisel 1981). Therefore, creative design largely "consists of problem definition and redefinition" (Rowe 1992).

Accordingly, if architecture studios, as Boyer and Mitgang insist, form "one of the best systems of learning and personal development that has been conceived" since they offer "unparalleled ways to combine creativity, practicality, and idealism" (Boyer and Mitgang 1996), first-year studios are particularly crucial in instilling good learning habits into beginning architecture students. In the schematic design loop presented by Zeisel, the spiraling movement towards the field of satisfactory

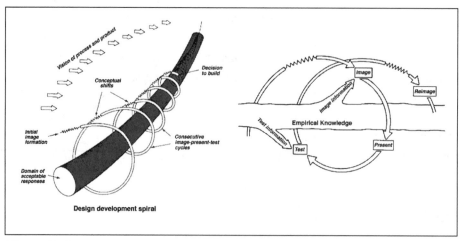

Figure 20.1 *Design development spiral and image-present-test cycle (John Zeisel, Inquiry by Design)*

solutions is accompanied by "consecutive image-present-test cycles" where "imaging" consists of going "beyond the information given" to form a "solution in principle." After the design problem's site and functional program have been analyzed and interpreted, "presenting" consists of the drawings and models necessary to communicate, evaluate, and improve images, and "testing" consists of the reflective and critical judgments that "look backward and forward simultaneously" and that prepare "the way for the next creative leap," or in other words, help define the problem further as well as refine the next images (Zeisel 1981).

In this cycle, the most likely entrance point where 3-D computer modeling would make a difference for first-year architecture students is "presenting." One had to be careful, however, not to spend too much time trying to modelize too nicely, which might impair the design loop and the time that it is allowed, especially with regards to the consecutive testing and re-imaging stages.

Expectations and Innovations

The primary goals of our experiment were to make use of 3-D modeling to help better develop the students' design abilities, to produce more extensive design propositions, and to make students initially competent in the use of computer modeling for design and presentation. It was expected that by having a better visualization and understanding of their own ideas, students would go on producing more complete presentations to check and improve these ideas sooner in the design process. The software we chose, Form Z, was selected because our school had already worked with it and because it makes computer-aided design and 3-D modeling easier and more efficient, while still being much more affordable than AutoCAD, and such.

Moreover, this experiment was also innovative in three ways:

- It involved a first-year compulsory design studio entitled Housing 1, which, according to our school curriculum, had to meet specific pedagogical objectives relating primarily to housing design and not to computer modeling proficiency.

- It addressed a group of 15 out of 56 students just entering their second semester and the same thematic studio, enrolled in the 3-D computer modeling group on a voluntary basis since they had to bring their own computers to school for the semester. (The school provided the software and a secure environment.) About half of the group had already dealt with computer drawing before, specifically with AutoCAD; the other half had very little acquaintance with computers, perhaps only word processing.

- As the studio's main advisor, I had no prior knowledge of 3-D computer modeling, a situation that caricaturizes many teachers of my generation and older. However, I could rely on two thesis students as teaching assistants, during studio hours only, for all the modeling and computing aspects of the project.

With regards to the design studio's importance in architecture education, the previous expectations and innovations implied that my pedagogical methods and exercises be adapted to the experiment and remain flexible, just in case something did not unfold the way I tried to anticipate. In terms of pedagogy, I relied on what I humbly like to call a "Socratic method." I encourage students to build on what they already know, then offer different means (readings, precedent analysis, etc.) to go on discovering by themselves what I wanted them to learn. It gave them confidence in the value of their own knowledge and their ability to go further.

I now realize that it also helped students feel at ease with 3-D computer modeling. They were given the opportunity to learn and use Form Z; but in the event that it would be too much to ask of first-year students, they would be able to rely on their drawing board at any time. In terms of design exercises, I tried to never isolate housing design and understanding from learning a new design and drawing tool. I thought that by doing both simultaneously, the image-present-test cycle would be enhanced, each element highlighting the others.

Pedagogical Exercises in Housing Design with 3–D Computer Modeling

According to the course syllabus for January 1998, the studio's main objective was to develop, by way of practical exercises, the understanding, synthesis, evaluation, and expression of ideas supporting the production of small housing projects. Emphasis was put on what a home represents and implications on housing design, (spatial definition, occupation, appropriation, and identification). The specificity of the studio, as the students already knew, was that a new design tool would be gradually integrated into the course's exercises. They were well aware, from the outset, that this tool was primarily meant to improve their decision-making.

Moreover, I constantly insisted that ideas were more important to me than nice drawings. Although this may seem paradoxical for a studio introducing a new design and drawing tool, I wanted to make sure that students would not fall prey to computer modeling feats at the expense of good thinking and good design. In this way, we will later see that home design, because of its many small spaces, was very interesting with Form Z, especially in terms of ambiance, sensory perception, and materiality.

The fourteen-week semester was divided into two distinct parts in terms of pedagogical objectives. The first was made of two exercises (two weeks each) intended to introduce students to the spirit of the studio as well as to 3-D computer modeling with Form Z. The second comprised two design exercises: the addition of a "haunted house" to a suburban bungalow (four weeks) and a small housing complex in Québec city (six weeks). Since the exercises were interconnected, the students were expected to accumulate and use knowledge developed from the previous work they had done. After a nice start and a few intermediary bumps, the last design exercise was proof of the amazing amount of work they had accomplished and of the very fine things they learned. The following is a sketch of their itinerary.

INTRODUCTORY EXERCISES: STAIRS AND PLACE

Although these exercises constituted a pretext for learning 3-D computer modeling, they were primarily meant to make students understand and illustrate some abstract notions qualifying home spaces. For the first one, entitled "Stairs," they had to read short essays by Bachelard (1983), Pérec (1974), and Tournier (1986) about the poetical and mysterious sides of stairs. Then, they had to react to those texts by producing a few hand sketches, which illustrated what they thought the idea of stairs

Figure 20.2 *Stairs (Form Z drawings: N-M. Paquin; J-N. Fortin; J. Bouchard)*

meant for a home space. They had to build this sketch/place with Form Z. If drawing a staircase happens to be a relatively easy first modeling exercise, some students' work shows how considerations about comfort, pleasure, and mystery have also been translated into and around the staircase. It was this translation that was graded, along with the effort they put into 3-D computer modeling.

The second exercise entitled "Place," dealt with the analysis, understanding, and judgment of the important ideas and the spatial and material elements that qualify a room, especially in terms of its relationship with outdoor space. Students were given 2-D drawings and photographs of an existing house and, after reading excerpts from von Meiss's *De la Forme au Lieu* (1986), they selected one room they thought best demonstrated the interest of the architect's choices in terms of form, sensory perceptions, materials, adjoining outdoor space, and sunlight. They then had to redraw or rebuild the room for themselves in order to know it and to highlight its characteristics as they continued with computer modeling. They were graded mainly on their understanding of the house and on how they shared this knowledge with fellow students. However, an interesting outcome of this exercise was that, because of their genuine enthusiasm for 3-D modeling, many students did the whole house. No wonder they were exhausted and that the next exercise suffered a little from "computer-sickness."

Figure 20.3 *Place (Form Z drawings: J. Bouchard and P. McCormack; C. Ferland and J-N. Fortin; N. Rioux and Y. Labelle)*

DESIGN EXERCISES: SUBURBAN "HAUNTED HOUSE" AND URBAN HOUSING BLOCK

With this first design exercise, it was expected that students would dwell on what they learned in the introductory exercises and that, exhaustion notwithstanding, computer modeling would slowly become an implicit habit for them. This exercise's scenario required them to design an addition to an existing house for Gaston Bachelard to live in, a chap with whom they had been acquainted before and they knew was fond of spatial poetry, reveries, and surprises. We concentrated on the haunted aspect of the house, which I thought 3-D computer modeling would help develop and present since this exercise's main challenge was to highlight the home's wonderful qualities with the simplest technical and material means.

Apparently, the four-week period devoted to this exercise was too short to develop home design and apply 3-D modeling at the same time. As a matter of fact, Form Z was used only for a few final presentation drawings and not as a design tool in accordance with the image-present-test cycle. Since it was part of our initial agreement that students were free to use whatever design tool with which they felt most comfortable, I did not mind having so few Form Z attempts and drawings because most of their design propositions met the exercise's pedagogical objectives.

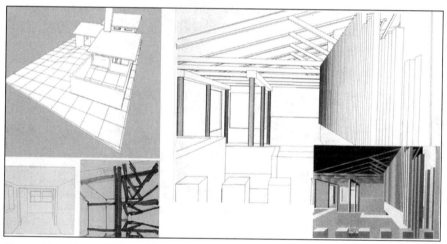

Figure 20.4 *Haunted house (Form Z drawings: N. Rioux; C. Ferland and J-N. Fortin)*

Considering what happened with the haunted house project, the students and I agreed that trying to start 3-D computer modeling as early as possible in the design process would be a reasonable goal for the last design exercise. I reminded them that, far from looking for the nicest modeling feats, I was rather hoping for drawings that would help them better test their design ideas, have more comprehensive discussions with me, and, therefore, react more quickly and efficiently when working.

Our last project of the semester consisted of designing an apartment complex in downtown Québec City. As they started with a one-week site and precedents analysis, students had to make volumetric propositions that took into account the neighboring environment and the sunshine that the site would eventually get, according to the most appropriate urban apartment types. We realized that Form Z was particularly helpful for such preliminary studies since students could measure early in the design process, and very precisely for that matter, the value and impact of their ideas.

Figure 20.5   *Site and precedents analysis (Form Z drawings: P. Duguay; D.S. Loya and B. Mareschal)*

Figure 20.6   *Urban housing block (Form Z drawings: P. Duguay and B. Mareschal)*

Figure 20.7   *Urban housing block (Form Z drawings: J. Bouchard and P. McCormack)*

Figure 20.8 *Urban housing block (Form Z drawings: B. Dubois and J. Gauthier)*

This was a very encouraging step for all of us, and most of the students continued refining, testing, and developing their ideas with the aid of Form Z for the remaining five weeks of the exercise. At the same time, they would add ideas from reading and interpreting Alexander's *A Pattern Language* (1977) to their project. As with the previous precedents analysis, which gave students tips for initiating the first design ideas, the "pattern"—a description of recurrent architectural problems and of their core solution—also helped adjust the design process. In conjunction with comprehensive Form Z illustrations, the spatial, perceptual, and material qualities of their design propositions are presented more fully. The quantity of drawings that can be produced rather quickly, once the core of the design has been modeled on the computer, is far from being negligible when it comes to sharing their ideas.

WHAT WE LEARNED FROM THIS EXPERIMENT

Considering all the things that beginning architecture students must learn in first-year studios, both in terms of design methods and presentation means, the addition of a new design and drawing tool constituted quite a challenge for all of us. As I got involved in such an experiment for the first time, not knowing anything about 3-D computer modeling myself, I decided to conduct this experimental studio much like I did my usual housing studios. I adapted the exercises to the new situation and hurried the pace a bit in order to leave time for the students to learn 3-D computer modeling with my teaching assistants.

I consider it a plus that I did not have to deal with the "computing" side of the studio myself. I believe that all technical questions or problems that may occur should be the responsibility of a third party (in this case my teaching assistants), so that discussions between teacher and students focus on design and on the pedagogical objectives we had to meet.

Looking back on this first experiment, and hopefully planning next year's episode, there are a few things that I would do differently. First, I realized that there was one introductory exercise too many. As the "Stairs" exercise showed, two weeks into the semester students already had a good hold on the mechanics of Form Z. Second, a readjustment is necessary between the other introductory exercise and the first design project. If the "Place" exercise was formative in terms of understanding the quality of a home, it was energy consuming and it also handicapped the time and enthusiasm devoted to the "Haunted House" project. Third, first-year architecture students need to be reminded repeatedly of the image-present-test cycle's importance for their projects. In this way, they are more likely to grasp the relevance of producing comprehensive drawings and models for a good development of their design abilities, and to start using 3-D computer modeling earlier in the design process.

The final design exercise demonstrated that 3-D computer modeling could be an efficient part of the design cycle as it is presented to beginning architecture students, simultaneously proving that first-year studios are not too early a stage to do so. Although its impact on the students' design abilities and on the projects' quality is still difficult to measure precisely, positive conclusions can be drawn from this experiment. Since 3-D computer modeling with Form Z gives them quicker and fuller visualization and understanding of their own design ideas, the students and I could travel their work thoroughly, and marvel at it. This was especially true with the numerous perspective drawings that presented what is too often left unseen in first-year students' work because it is so time-consuming to do on the drawing board.

Ambiance and materiality could be tested more efficiently, not just guessed at, making it possible for students to learn and teach themselves better and to share this knowledge more easily. Moreover, the students' improvement with 3-D computer modeling over the semester was impressive; their enthusiasm was constant and surely matched by Form Z's reassuring ability to truly represent what they had in mind, especially for those having more difficulty with hand-drawing. I am confident that their next studio experience will be even more rewarding because the last exercise made them realize the benefits of using 3-D modeling at the early stages of the design cycle.

All in all, the experiment went smoothly. The ambiance of the studio was amazingly happy in spite of all the work the students had to accomplish and of which we all were very proud. Since I chose never to grade their computer proficiency or the niceness of the drawings they produced that way, but rather the general quality of their projects (which Form Z enhanced, in most cases), I believe they felt more comfortable with the concept from the start. In general, they appreciated being given the opportunity to learn 3-D computer modeling this early in their undergraduate studies and as a part of learning the design process. All the work presented here is proof of their success.

References

Alexander, Christopher (1977) *A Pattern Language: Towns, Buildings, and Construction.* New York: Oxford University Press.

Bachelard, Gaston (1983) "La Maison de la Cave au Grenier," in *La Poétique de l'Espace.* Paris: Presses Universitaires de France.

Boyer, Ernest L. and Lee D. Mitgang (1996) *Building Community. A New Future for Architecture Education and Practice.* Princeton: The Carnegie Foundation for the Advancement of Teaching.

Pérec, Georges (1974) "Escaliers," in *Espèces d'Espaces.* Paris: Galilée.

Rowe, Peter G. (1992) *Design Thinking.* Cambridge: The MIT Press.

Tournier, Michel (1986) "L'Esprit de l'Escalier," in *Petites Proses.* Paris: Gallimard.

von Meiss, Pierre (1986) *De la Forme au Lieu. Une Introduction à l'Étude de l'Architecture.* Lausanne: Presses Polytechniques et Universitaires Romandes.

Zeisel, John (1981) *Inquiry by Design. Tools for Environment-Behaviour Research.* Cambridge: Cambridge University Press.

Endnotes

The students who took part in the experiment were Julie Bouchard, Stéphanie Caron, Bryan Dubois, Pierre Duguay, Charles Ferland, Jean-Nicolas Fortin, Joanne Gauthier, Annie Géhin, Yan Labelle, Fabienne Lamontagne, Daniel S. Loya, Benoît Mareschal, Philippe McCormack, Nicolas-Mallik Paquin, and Nicolas Rioux. My teaching assistants were Geneviève Giguère and Romain Chauvelot. I thank them all very sincerely for their enthusiasm, especially with regards to the impressive amount of time and work they put into this pilot project.

CHAPTER 21

Mapping, Discussing, and Crossing Borders: Computer Technology as a Mode of Inquiry in Three Humanities Courses

Part 1

"I Read the News Today, Oh Boy!": Spanish Newspaper Web Sites Link Students to 20th Century Spain

Dawn Smith-Sherwood
Assistant Professor of Spanish

Jacksonville University
Jacksonville, FL

INTRODUCTION

Spanish language teachers have long employed *realia* in their classrooms, enabling students to see connections between course content and the real world. Exercises involving canceled train tickets and colorful grocery store fliers gathered abroad not only demonstrate grammar and vocabulary in action, but also concretize the people who use Spanish every day. An often-utilized form of *realia* is the Spanish-language newspaper. Its official, "black and white" authority makes tangible the existence of another culture and immediately lends legitimacy to the whole foreign language teaching enterprise. At the elementary and intermediate levels, students complete activities related to weather forecasts, movie theater timetables, and horoscopes. At the advanced level, students read and interpret articles, honing their comprehension skills and augmenting their cultural awareness.

Unfortunately, practical problems related to edition, time, and cost frustrate the most effective use of Spanish-language newspapers in the classroom. Condensed weekly international editions are available by mail, but many interesting items (e.g., weather reports, local advertisements, and event schedules) are omitted for reasons of space. By the time the paper arrives, the news is old, diminishing the students' sense of connection to Spanish-speaking people abroad. While edition and time concerns constrain utility, cost sometimes eliminates it altogether. Under pressure to cut budgets, departments and libraries are apt to eliminate costly Spanish-language publications. The remnants of once enriching *realia*, yellowing stacks of El Pause issues, are a common sight in the offices and lounges of today's Spanish language teachers.

Enter the age of "virtual really." The recent creation of Spanish-language newspaper Web sites has closed the gaps-spatial, temporal, economic-previously inherent in the classroom use of their hard copy counterparts. Teachers and students alike can now approach the current edition on the same day, and as cheaply as their co-readers abroad. In spring 1998, I put the potential merits of virtual *realia* to the test in an advanced-level college course concerning Spain's turbulent last century (1898-1998). Organizing and assigning up-to-date news, culture, opinion, and society pieces via my home page (http://junix.ju.edu/UserWebPages/dsmiths), I encouraged students to discover in Spain's present the threads sewn in its past.

JACKSONVILLE UNIVERSITY: SPANISH 321– CONTEMPORARY HISPANIC ISSUES

An English translation of the course description reads:

> From the loss of the last American colonies to the twentieth anniversary of the 1978 Constitution, "Crisis and Calm: The Last Spanish Century, 1898-1998" will examine "one hundred years of intensity" through history, literature, and art. The course will consider six historic-cultural periods: the Generation of 1898; the Generation of 1927; the Second Republic (1931-1939); the Civil War (1936-1939); the Franco Dictatorship (1939-1975), and Transition and Democracy (1975-present), while investigating echoes of the past in the journalism of the present. The readings (traditional and by Internet), discussions in

class, lectures, and movies will serve the students as fodder to gain a
deeper understanding of Spain's complex twentieth century.

The creation of this course enacted a conviction that students would more
fully apprehend contemporary Spain through simultaneous study of past and
present. The teaching of this course demanded a pedagogical act of faith: Though
odds were favorable that the Spanish media would commemorate this significant
year in opinion pieces, scouring each week's online dailies for appropriate news
material could prove nerve-wracking. What if the news did not synchronize with
historical course content?

From President José María Aznar's renewed coalition with the Catalan President
Jordi Pujol to the most recent ETA attack, the news from Spain linked classroom
history with contemporary events as they were happening. At the beginning of the
semester, *El País* (www.elpais.es), Spain's largest-circulation daily, began a series
of articles regarding Spain's adoption within the year of the euro as the new unit of
currency. The most concrete evidence of Spain's participation in the European
Union to date induced an exploration of the lingering tension between competing
desires of *españolización de Europa* and *europeización de España*. For students, the
issues of 1998 linked with those that troubled the writers of the Generation of 1898,
the period with which the course began. Students observed how these issues continued to surface throughout the twentieth century as Spain oscillated between dic-
tatorships and attempts at republic, and they perceived issues central to the
Generation of 1898 in Spain's contemporary struggle to secure for itself an identity, both separate from and in relation to Europe.

Articles taken from the online edition of *El País's* culture, opinion, and society sections were also valuable complements to traditional course materials.
During the segment of the course devoted to the Generation of 1927, students
studied the drama and poetry of Federico García Lorca, a leading member of the
generation and the most internationally familiar Spanish author of this century.
Lorca's figure haunts any study of twentieth-century Spain due to his birth in the
significant year of 1898 and his untimely death, murdered by Francoist forces at
the beginning of the Civil War in 1936. The 1998 centennial of Lorca's birth
occasioned noteworthy tributes in the culture pages of *El País*. Musicians from
around the world, including the lead singer of the hit group Blur, were recording
an album of songs that utilized Lorca's poetry as lyrics. A Spanish theater troupe
was staging a parody of a lesser-known Lorca play. Reading these current
articles provided students with a chance to recognize Lorca's importance to

contemporary Spanish and world cultures while approaching and discussing his literature, written more than sixty years ago.

Reflective essays regarding the current *fin de siglo* filled the opinion pages of El País in the first months of 1998. The most useful of these for my own course appeared in the January 22 issue. Antonio Elorza's "Crisis de Imperio" contemplated the centennial's arrival, cataloging significant Spanish events in the years through 1998. This essay provided a brief panorama of 500 years of Spanish history, from the death of Torquemada, the infamous inquisitor, in 1498 to the present day. The *hilo rojo*, the historical trail of blood that Elorza traced, led twentieth century students to investigate events surrounding the Spanish Civil War and its aftermath.

On a lighter note, the entertaining society pages of *El País* also offered educational opportunities. The Spanish press's coverage of the dashing Prince Felipe's thirtieth birthday led to class discussion of the role in the Spanish government of the prince and King Juan Carlos I, his father, as outlined by the Constitution of 1978. Student interest in the lives of royals promoted inquiry into the historical role of the Spanish monarch and its current function within "parliamentary monarchy," a concept unfamiliar to students in the United States.

PROBLEMS WITH USING THE WEB IN COURSES

As with any new technology, incorporating Web-based readings into a course requires teachers and students to be on the watch for potential pitfalls. The complications inherent in the utilization of daily Web texts may occur at the level of the online publication itself or the access technology provided by the institution. Difficulties that stem from the publications themselves include the unpredictability of broken links. Students may not be able to complete an assignment on time if they are unable to access pertinent material. A related situation may arise due to a lack of archiving at the dailies' Web sites. Most save only one week's worth of issues, so students who fail to complete an assignment on time may lose access to the material forever. To avoid both of these situations, the instructor may save copies of assigned material as backup files on the local server. However, this solution carries the disadvantage that students accessing the backup copy of the article miss the opportunity to browse other articles in the daily.

At the level of the institution, potential problems include the risk of system failure. A too frequent cry of "The server is down!" will quickly dampen everyone's enthusiasm for the use of electronic media in the classroom. Insuring a sufficient number of computer stations at which students may gain access to the Web may also complicate the delivery of Web-intensive courses at some institutions. Students who rely on the institution for access to the Web will be frustrated if there are not enough terminals for their use. As institutions update their hardware and become increasingly wired, these issues should be resolved.

One of the difficulties facing instructors who rely on electronic media has already been mentioned above: One cannot anticipate how the news will relate to the course outline. This problem may also affect future incarnations of a particularly successful course. For example, while there has been much reflection in the 1998 press on the last Spanish century, there is no guarantee that next year's coverage will be as pertinent to a similar course. An additional complication lies in the need for increased class preparation time in order to review electronic sources in search of appropriate course material as well as maintaining the instructor's home page. Also, instructors requiring technical support to complete such maintenance must allow enough time for consulting with institutional technology staff.

There are also the low-tech problems, including old-fashioned student resistance to the use of electronic media. The perceived "softness" of electronic texts, as compared to traditional, "hard" texts causes some students to take Web-based assignments less seriously than their paper forms. "I didn't get a chance to do the online reading." speaks volumes about the Web's perceived supplementarity. If Web resources are to be used as primary rather than secondary materials for a course, instructors must take specific steps to dispel students' "extra credit" perceptions. Be quite specific regarding the assignment to be completed using the electronic text. Asking students to read an article is not enough. Students may be required to come to class with several questions that they feel their classmates should be able to answer from having read the text. Students may be asked to present an oral summary of the article to their classmates. The completion of discrete tasks related to Web texts improves student perceptions of their seriousness and worth, and translates into more effective use of these texts.

Instructors interested in incorporating electronic texts into their courses should assess their institution's system reliability and their student's ease of access. They must also be willing to undertake the time-consuming responsibility of home page construction and maintenance, or alternatively, have access to adequate institutional support. Be prepared to be flexible. Adapt and integrate virtual realia as it arrives.

FUTURE PLANS FOR TEACHING USING THE WEB

In future incarnations of this course, I plan to increase the use of virtual *realia*. The variety of Spanish newspaper resources currently available through the Internet will permit the review of periodicals from across the country's political spectrum. Electronic editions of three of Spain's major dailies—*El País*, *ABC* (http://www. abc.es), and *El Mundo* (http://www.el-mundo.es), from center, right, and left respectively-offer a range of ideological perspectives that reflect conflicting viewpoints of twentieth-century Spain that have persisted through the Civil War, the Franco dictatorship, and 20 years of parliamentary monarchy. Comparison of articles from across the well-defined Spanish political spectrum will contribute additionally to the sense of energy that informed the original course. Also, the unpredictability inherent in teaching the past in the present through daily Web sources will be balanced by the use of sites whose content is relatively static, (e.g., Tourist Office of Spain [http://www.okspain.org], Museo del Prado [museoprado.mcu.es], and Generalitat de Catalunya [http://www.gencat.es]).

In addition to incorporating supplemental Web sources that provide historic-cultural information, I hope to have the opportunity to use a fully computerized classroom when next teaching this course. This will allow Web texts to be available in their electronic form during class meetings rather than having to rely exclusively on paper copies or students accessing material outside of the classroom. In an effort to encourage comfort and reliance on electronic texts, students will be made responsible for searching out and identifying some of the pertinent Web materials.

BENEFITS OF USING THE WEB IN COURSES

Providing direct, immediate, and free access to such Spanish dailies as *El País*, *ABC*, and *El Mundo*, the Web links students to twentieth-century Spain. In teaching this version of contemporary Hispanic issues, I found that the incorporation of electronic news articles facilitated a productive inquiry into the ways in which Spain's past continues to impact its present. Electronic editions of major dailies are currently available from many other countries providing excellent resources for similar foreign language courses. (For an index of foreign press online, see "The Daily News" at www.middlebury. edu/~gferguso/ news.html). Similarly, teachers of advanced-level composition and conversation courses will find these resources useful for providing students with quality prose models and an ample supply of topics on which to write or discuss.

Even at the elementary and intermediate levels, the Web is proving useful; current editions of several Spanish language textbooks include Web exercises for students to complete, immediately challenging and augmenting their language skills as well as their cultural awareness. (Holt, Rinehart, and Winston's *Dímelo Tú* and McGraw Hill's *Al Corriente* are noteworthy examples.) For all instructors who strive to open their courses to the world, the Web provides an excellent resource, keeping teaching current and course content fresh, protecting students from instructors' yellowing note cards and foreign language teachers from stacks of yellowing international newspapers.

Part II

Mapping Out Borders:
Computer Cartography as a Mode of Inquiry
in a History Course on Absolutism in France

John D. Garrigus
Associate Professor of History

Jacksonville University
Jacksonville, FL

INTRODUCTION

Students on our campus see history as a discipline for the technologically challenged. Members of our naval ROTC unit who can't pass Physics II, for example, choose a history degree more often than any other degree. The experiment narrated in this section, therefore, began as a coldly calculated attempt to force apprentice historians to use computers to process something besides words. But the success of the attempt created a host of new historical questions to explore in a class that focuses on Renaissance and Baroque France.

Perhaps because our department has no historians who do qualitative research, many of our students are attached to a caricature of the discipline, which holds that historians work exclusively with texts, as researchers and communicators. Beyond word processors and videotape players, our students rarely see their history professors using late twentieth-century hardware in their work. Many seem to assume, therefore, that majoring in history will provide them with few "marketable skills" beyond the proficiency they acquire in written and oral analysis.

THE INCORPORATION OF COMPUTERIZED MAPS INTO A HISTORY COURSE

I knew this to be false, but my own set of computerized research tools are textual databases that took months to construct and are hardly appropriate for student projects. However, the experience of creating my own maps for a series of articles awakened me to the relative ease with which non-cartographers could produce professional-looking historical maps. I decided to teach students how to use computer graphics in an upper-level class called "France and the Sun King," a political and cultural history of the years from 1515 to 1715. The experiment, I hoped, would accomplish four goals:

1. Teach students French history more effectively, building their familiarity with spatial relationships in the events of the period.

2. Increase students' comfort with and interest in computers as tools for analysis and communication.

3. Engage students in a mode of research and exposition that would complement the writing and discussion skills that are at the core of the discipline.

4. Raise student awareness of maps as sources and as documents to be analyzed critically.

By accomplishing these objectives and by displaying students' completed maps on my Web site, which is widely used by my general education courses, I hoped to attract more and better students to this sometimes under-subscribed class.[1]

The integration of computer graphics into this history course (HY326) in fall 1997 was just one part of a complete revision of this course. Rather than present the history of France as one chronological narrative, I organized the semester into six separate units, each presented as its own story in readings, lectures, and research assignments. In each unit, the class examined the rise of monarchical state power vis-à-vis topics like the evolution of social classes, the struggle between Protestants and Catholics within the kingdom, and the conquest of France's Atlantic empire. In oral mid-term and final exams, students displayed their knowledge and ability to integrate these themes in a discussion of the power of Louis XIV, France's self-titled "Sun King." Moreover, for each unit, I created a list of research questions. Students worked cooperatively in pairs, with very specific individual responsibilities, to research and write two formal papers on questions they had chosen from the list.

The computer graphics assignment was not one of these collaborative research projects, but an entirely separate element of the course. In addition to their research papers, I required students to produce an accurate and well-designed map illustrating some event, situation, or process in French history that we studied during the course of the semester, and I again provided a list of possible topics. I set aside four of our 28 class meetings (each lasting 80 minutes) to teach the computer graphics program and the basics of map creation. In one of these meetings, students worked through the software's own tutorial, and in two other sessions they followed elaborate guides I had created, based on my own experience with the program. I reserved one final session for troubleshooting and fine-tuning students' completed maps.

THE SOFTWARE

The software I selected for this experiment was Macromedia's Freehand 7. This program offered considerable advantages. I had used earlier versions for my own scholarly projects over several years. Macromedia offered an attractive educational license ($600 for 10 copies) and a colleague in the art department contributed about half of the funds needed for this package. Freehand had emerged early in the history of graphical computing as an industry standard on the Macintosh platform, and though we would be using the Windows 95 version, it was widely reviewed as one of the top programs in the professional graphics category.

Because I hoped to develop a collection of student cartography I could easily show on the Internet, I was attracted to Freehand's ability to convert its images into Shockwave files. Shockwave is free software that works within a Web browser (a "plug-in"), allowing viewers to manipulate pictures on their screens. This feature seemed ideal for displaying the complex maps I hoped students would eventually produce. In fact, many professional cartographers use Freehand for their work. Avenza's MAPublisher 3 is a software package that allows a Freehand graphics file to be published on the Internet as an interactive digital map that can be altered to illustrate underlying data sets. Though this was far beyond my modest goals for the class, the options for expansion were attractive.[2]

It was also important that Freehand be a "vector-based" drawing program. This label means the software uses mathematical formulae to generate lines and shapes, so it can easily magnify or reduce images without compromising their quality. Many other drawing packages, like the Paint program that comes with Windows 95, are "raster-based," meaning they create lines and shapes pixel by pixel on the screen. This process is very intuitive but the resulting images lose their clarity when their dimensions are changed, as is frequently necessary when maps are to be squeezed onto a Web page.

One final advantage of Freehand for this application is its "layers" feature, which allowed my students to use the program as a kind of electronic tracing paper. I taught students how to save public domain map images from the Internet onto their computers. They then imported these graphics files into Freehand, placing the image on a background layer. They traced key features of their own maps in another, transparent layer over the public domain original and deleted the background layer once they had duplicated the details they needed. For the rest of their map creations, the design of the legends and symbols, and the interpretation of historical events, students were on their own, with advice from me.[3]

From the first of our four sessions in a campus computer lab, it was obvious that students had very different aptitudes for this kind of work and that my role as a teacher would be quite different from what it usually is. Because we only had 10 copies of the software, I divided the 16 students into eight pairs. By doing this, each has a partner with approximately the same level of computer experience. All but one of the students had completed our campus's required "Introduction to Computing Systems" course, but some students nevertheless required extensive coaching. Because I had written tutorials with detailed instructions, students mostly needed help correcting minor errors they had committed in the Windows 95 interface.

For a few, however, the tutorial method itself was the problem. One typically high-achieving history and studio art double major who is a talented painter and sculptor was so intrigued by the graphic capabilities of the program that she didn't follow the tutorials closely. Without this structure, however, she quickly grew frustrated with the software, whose drawing tools have little in common with a brush and canvas. Such students need a very different kind of instruction, one that shows them the program's different modes and then gives them time to experiment with their own images, rather than leading them through a narrow recipe.

THE IMPORTANCE OF STUDENT ACCESS

As this episode illustrates, students' ability to adjust to the personality of the software was a critical element of their success with the assignment. This adjustment takes time and, ideally, students should have ready access to the program, either on their personal machines or in a wide assortment of labs across campus. Those students who were able to return to the computer lab and practice between classes moved ahead rapidly, while those who did not spent class time relearning basic techniques.

Because of the importance of student access, one possibility for a future offering of the course is to trade Freehand's Shockwave capabilities for a raster-based shareware program like Lview or Paintshop Pro, and recommend that all students acquire their own copies for home use. One of the best students in the class, in fact, insisted on using PaintShop Pro for her map. She produced a very attractive final product, in part because she was able to work at home. However, because of the size and format she chose, her final image could not be easily altered to post on the Internet.[4]

SUCCESS WAS ACCOMPLISHED

Despite problems with tutorials, software, and other elements of the map assignment, the cartography section of this experimental class was a success. Judging by their final exams and completed map projects, students did acquire a greater familiarity with the spatial elements of French history. Nearly all achieved a greater level of basic ability and self-confidence with computers. And, though I have been using structured cooperative learning techniques in class for several semesters now without computers, this was the first semester I felt students were working effectively in teams. I attribute this success to the computer component of the course. For, as the class progressed through the tutorials, the more advanced students became coaches for the others, demonstrating techniques, helping classmates solve problems, and working together outside of class. Using this non-traditional technology seemed to dramatically affect the class mood. The inevitable glitches in hardware and software transformed me from instructor into partner, as students and I worked together to solve problems in an area where I knew only slightly more than they did. Finally, students' imaginations were stimulated by their new ability to make professional-looking illustrations. Three of the students came to me after the class ended to get my advice on a map they were making for another class, using our techniques. One student is using her map-making skills for a special Internet project.

FUTURE DIRECTIONS OF THIS COURSE

This success has convinced me that the next time I offer this course, cartography will again be a major component. The challenge in future versions of the class is not just to avoid the technical frustrations of the initial semester, but to structure the assignment so that the technology is more transparent and the historical material and controversies more salient. Rather than separate the map assignment from the rest of the course because of the technology, I will instead make the cartographic inventions and controversies of early modern Europe one of the six central sections of the course. This will not be difficult to do, since this was the period when modern western mapmaking was invented.[5]

The theme of the course is the rise of the central state in Baroque France, and maps were an important symbol of Louis XIV's power. More time spent analyzing maps drawn in sixteenth- and seventeenth-century France and understanding their political and cultural uses will expand students' awareness of the problematic relationship between maps and historical "truth."[6]

Discussing and dissecting these maps will also lead students to focus on the elements of map construction that make an effective illustration. The students' final projects show that I need to make design issues a more prominent part of the course. For example, I will take previous student maps, remove their authors' names, and post a critique of their work on the Web for future classes.[7]

Students became so engrossed in the computer technology that they lost sight of the original assignment, which was to produce a map illustrating an historical process, event, or situation. Although I had established a list of possible map topics, many had difficulty producing data to support the map topic that captured their interest. Under the pressure of a deadline, some were reduced to paraphrasing maps they found in secondary sources, though always with an original twist. In order for the map assignment to reach its full potential, history students in the computer lab—like their counterparts in biology, physics, or sociology—need to be presented with raw data if they are to focus their energies on interpreting the past and presenting their results.

Endnotes

1. The syllabus, tutorials, and some student maps from this class can be found in the HY326 link from http://www.ju.edu/user/~jgarrig.

2. See http://www.macromedia.com/freehand and http://www.avenza.com for more information.

3. My instructions to students on how to trace Internet maps in Freehand 7 can be found attached to the syllabus for HY326 on my home page at http://www.ju.edu/user/~jgarrig. An outstanding source of United States government maps is the Perry-Castañeda Map Library at the University of Texas at http://www.lib.utexas.edu/Libs/PCL/Map_collection/ Map_collection. html. The *CIA World Factbook* provides simpler country outlines in its maps at http://www.odci.gov/cia/publications/factbook. An excellent list of map server and reference links can be found at http://www.cgrer.uiowa.edu/servers/servers_references.html.

4. Evaluation copies of both PaintShop Pro (http://www.jasc.com) and Lview2 (http://www. lview2.com) are available for downloading. After a free 30-day evaluation period, however, users must buy a CD-ROM version of the program. Lview2 is available for $20 a copy when more than 2 copies are purchased. PaintShop Pro 5, a more powerful product, costs $99 a copy, according to the online literature. My online forays in the shareware world have not found a vector-based shareware program that has the features necessary for drawing maps, but I do not claim to have completely surveyed the field. Two popular shareware sites are http://www. shareware.com and http://www.cnet.com under "downloads."

5. John Noble Wilford, *The Mapmakers: The Story of the Great Pioneers in Cartography from Antiquity to the Space Age* (Vintage, 1982).

6. Tom Conley, *The Self-Made Map: Cartographic Writing in Early Modern France*, (University of Minnesota Press, 1996); Marcelo Escolar, "Exploration, Cartography and the Modernization of State Power," *International Social Science Journal* 49 (March 1997): 55-75; Gregory H. Nobles, "Straight Lines and Stability: Mapping the Political Order of the Anglo-American frontier," *Journal of American History* 80 (June 1993): 9-35; David Buisseret, ed., *Monarchs, Ministers and Maps: The Emergence of Cartography as a Tool of Government in Early Modern Europe* (University of Chicago Press, 1992); Denis Wood, "The Power of Maps," *Scientific American* 268 (May 1993): 88-93; Walter D. Mignolo, "Misunderstanding and Colonization: The Reconfiguration of Memory and Space," *South Atlantic Quarterly* 92 (Spring 1993): 209-60; Norman J.W. Thrower, *Maps and Civilization: Cartography in Culture and Society* (Chicago, 1996); J.G. Harley, "The New History of Cartography," UNESCO Courier, 1991 (June 1991): 10-15.

7. A survey of cartographic techniques and approaches can be found in Mark Monmonier, *Mapping It Out: Expository Cartography for the Humanities and Social Sciences.* (University of Chicago Press, 1993) and Mark Monmonier, *How to Lie with Maps* (University of Chicago Press, 1996); see also Jeremy Black, *Maps and History: Constructing Images of the Past* (Yale, 1997).

Part III

Mexican Connections:
The Use of Class Web Bulletin Boards in
Researching and Discussing Issues of
Contemporary Mexico

Douglas Hazzard
Associate Professor of Spanish

Jacksonville University
Jacksonville, FL

INTRODUCTION

To address the efficacy of the World Wide Web as a tool allowing students to investigate contemporary issues—in a course (taught in English) considering the question "What is Mexico?"—I will describe this interdisciplinary honors course and present the rationale behind the use of this technology for student research and online discussion. I will describe student assignments and discuss students' perspectives on the value of this course component. Beyond the specific course, I will address issues of wider interest to faculty, such as teaching critical approaches to Web materials, encouraging cross-disciplinary use of language skills, and fostering an intellectual community in an online environment that some argue encourages disconnectedness and shortens attention spans. Also, I will assess the future uses of this technology in the humanities and address the problem of technological support and faculty motivation.

The use of technology as a substantive component of a humanities course certainly acknowledges its ever-growing importance in today's academy and in the world outside the classroom—the world in which our students will be living and working. Students have come to consider the technological component as integral, even commonplace, in all learning endeavors; and in this new academic order the humanities classroom, putatively, should be no different. As a middle-aged professor who has reluctantly become passably proficient with computers, widely considered a neo-Luddite on my campus, I have two significant problems with the privileging of technology in the classroom, both of which reflect a discomfort with my students' notion that technological progress is inevitably positive and so to be welcomed. First, I remain unconvinced that what we might label the de-centered (the postmodern) classroom enhances learning. Second—and this self-conscious admonition explicitly refers to the interdisciplinary honors course in which I use the World Wide Web most, a course whose general objective is to familiarize the students with Mexico through its history and culture—I am bothered by the suggestion implicit in the very use of technology in the course that Mexico will not have "developed" until it is more like we are. It will regrettably remain wholly other until it fulfills United States notions of progress toward a "First World" of technological literacy. Is it contradictory to assert, as I do, that Mexico's power, its identity, its very vocation is culture, while I privilege technological advances as methods of acquiring knowledge of this culture in the classroom? Perhaps not; but it is somehow an inconsistency that to me lies murkily at the core of this particular course and of the contemporary classroom in general.

COURSE DESCRIPTION

This contemporary world studies course, entitled simply "Mexico," is structured around three indistinct geographic regions, which for me serve to represent the three different Mexicos, whose histories I present, not always chronologically, from pre-Colombian times to the present.[8] Thematically, the *centro* is presented as the ideological foundation and engine of the modern Mexican state. The *norte* is presented as the literal and metaphorical border/*frontera*, with its area-specific economic and social identities. The *sur* is presented as the paradoxical "past/future" Mexico, the arguably authentic "nation" with its rich and complicated indigenous heritage, its almost stereotypically troubled present, and its interesting, if distant possibilities, for racial and economic reconciliation.

We read three novels in the course, see three films, and read and view numerous other videos and texts, including short pieces in a course pack and the longer collection of Subcomandante Marco's communiques from the base of the EZLN in the jungles of Chiapas.[9] There is no history text for the class, as I want pertinent facts about events and people to be gathered from multiple sources, rather than through the authority of a class textbook. Class time is organized around a combination of lectures and class discussions, always beginning in small groups and finishing as a whole class. Half of a student's grade is earned through three take-home essay exams, each corresponding to an area and a third of the course, the last counting a bit more and being cumulative. The other half is comprised of graded participation in class discussion and Web work.

It is the number and variety of sources and lack of one authoritative text (other than the professor, who nevertheless attempts to be just one of many sources) that is most innovative and controversial about this approach to teaching about Mexico (or any subject). My intent is to provide the students with a considered body of information, just as an history text might do, but also to give the students something more intangible: a sense of Mexico, a feel for the country and its peoples. Ideally, they get the facts, the foundation needed to interpret what they are learning, but they also get much more: a steadily deepening awareness of the rich composite that is Mexico, provided to them by many voices in many settings. And in this aspect, primarily, the Web becomes useful and beneficial.

I give the students five general questions that they are to explore through the Internet. These five questions cover contemporary issues but those that also have their roots in Mexico's very essence; and they are means by which students can dialogue with classmates and others about course-related issues that also happen to be very topical and pertinent to their own lives.[10] The five questions, each to be considered during a three-week period, relate to NAFTA, immigration, the drug trade, the PRI (the party that has governed Mexico since its revolution), and Catholicism.

During the first week, students are to post at least one pertinent, substantive link on the topic and accompany the posting with an analytical comment. The second week is devoted to developing a discussion on the links and comments posted by classmates. The third, concluding week is used to summarize the most promising discussion while relating it to the material being presented and developed elsewhere through readings, or lectures, or class discussions. The class is divided into three groups, each with its own Web bulletin board, so that each group has fewer, and presumably more pertinent, postings to consider. The quality of the discussion is also enhanced and sharpened due to the smaller number. Bilingual students, or those with an appropriate proficiency in Spanish, are given extra credit for posting links in Spanish and translating these for the group. This greatly improves the variety of perspectives in the pool of links, giving students access to points of view otherwise unavailable to them, besides offering language students an opportunity to display and hone their skills.

Students are evaluated weekly on the quality of the links and of their posted comments, and on their efficient and creative ability to interact informatively with each other and with the themes of the course as a whole. In defining what I mean by "quality," I ask students to develop a critical point of view in relation to the links and the comments. Do they, for example, vary the sources for their links and spend some time conducting a search and evaluation before posting a link? And when considering the links and comments of others, do they show critical thinking, or are their remarks superficial or unconsidered? And again, a primary component of this critical point of view is the informed and ever-developing ability to connect what students are learning on the Web with what they are learning from the other course "information sources," to use an infelicitous, but perhaps suitably technological term. I do not "appear" on the bulletin board, except to post the next question.[11] Students receive a Web grade every three weeks, along with a short written paragraph commenting on the quality of their work and, in effect, explaining the given grade.

WEB-RELATED DIFFICULTIES WITH THE COURSE

The practical difficulties encountered with the Web component of the course are fairly predictable. As my university continues to upgrade its facilities, there will continue to be inevitable shutdowns of the system. These occurred particularly at the beginning of the semester and complicated what was already the difficult task of exposing students to new ways of doing things. However, once we solved the technical and methodological troubles in the program, it ran very smoothly. Once all the students had ready access to the Internet, understood what I was requiring, and learned to organize their schedules to enable fulfillment of these new types of weekly tasks, the semester was two weeks old, so some minor adjustments in the time-line for considering the first two questions were made. I will be able to avoid these problems next time I teach the course, and I will be much better prepared to anticipate difficulties.

The more substantive concerns, those that are not technical or organizational in nature, may be equally easy to predict, but are more troubling...and interesting. Although the students generally liked the Web assignments and considered them informative, beneficial, and even fun, I found that their comments rarely fit well or appropriately within the general themes of the course. They were usually unable to connect critical information or opinion gleaned from the Internet with that learned elsewhere. And they were unable to evaluate the worth of the material received through the Internet, specifically by understanding—and assimilating that knowledge—the point of view informing the material.

These inabilities are particularly frustrating because, even though the students have been prepared and encouraged to consider critically what they are discovering, they seem to accept blindly what they learn from the Internet. However, this lazy tendency to respect all sources equally is one that we regularly deal with in the classroom, most often with students during their first or second year in college. There is little reason why a student should look upon the Internet as source any differently, with any more judgment, and it serves to remind us of the need to teach critical evaluation skills for the educated interpretation of all texts, including those on the Internet.

CONCLUSION

My experience from this course taught me what should perhaps have been obvious all along, but I too became victim of the convenience and novelty of the Internet. It can be a wonderful source. It certainly provides almost immediate access to texts that otherwise would be completely unavailable, particularly to the undergraduate, or take weeks, or even years to become widely available. If the Internet is to be used in the classroom-particularly by those professors like myself who are somewhat insecure as they negotiate the Internet, and almost automatically assume that their students are much more acclimated to cyberspace than they—then the professor must exercise his/her profession and teach students to read critically in all spaces.

Endnotes

8. At my university, this area of study resides in the College of Arts and Sciences and comprises courses taught by faculty from the humanities and social sciences divisions. The series of courses in contemporary world studies (CWS) was originally conceived to offer Bachelor of Science students an alternative to taking language courses. Since 1996,

Jacksonville University has had a language requirement for all majors, and CWS courses are now offered as electives in area studies.

9. The novels are Cormac McCarthy's *All the Pretty Horses*, Gloria Anzaldúa's *Borderlands/ Fronteras*, and Rosario Castellanos's *Nine Guardians*. The films are *El Norte, Like Water for Chocolate*, and *Lone Star*. The Marcos text is *Shadows of Tender Fury*.

10. An objective of this course is to expose these connections, which are not always obvious. Although most students enroll in the course because of an interest in Mexico, very few understand the nature of United States links to a culture, a people, an economy that can seem so foreign.

11. This was a much-considered decision, which perhaps requires its own lengthy justification. In short, I believed that my intercession would affect the discussion too much, overwhelming whatever benefits would be derived from what would presumably be my judicious, better informed comments. I now believe this decision was a mistake. I will still keep my comments short and to a minimum, but the next time I teach the course I plan on guiding the Web discussions more, modeling my mediation on the manner in which I might lead a discussion in the classroom.

Chapter 22

The Efficacy of an Electronic Modularized Course for Preservice Teachers: The Multicultural Education Course

Nelly Ukpokodu
Assistant Professor of Education
University of Missouri—Kansas City
Kansas City, MO

Education keeps pace with the life and growth of the community, and is altered both by changes imposed on it from without and by transformations in its internal structure and intellectual development. And since the basis of education is a general consciousness of the values that govern life, its history is affected by changes in the values current within the community.

—Werner Jaeger (1945)

As we rapidly approach the twenty-first century, it is clear that we are well into a new postindustrial and postmodern era driven by technology and information networks, among other things. Futurists warn that succeeding in an information-technology age will require new skills such as system thinking, experimental thinking, and collaboration (Wirth 1992). United States businesses are concerned that unless there is a shift in paradigm, the chances of surviving in an electronic, globally integrated market economy is very slim.

Today's and future workplace requirements are shifting to incorporate skills such as problem solving, team skills, learning to learn, and the capacity to interact, process, and apply information. Therefore, it is imperative that educational institutions bear a great responsibility for preparing all students to become successful in an information-based society. Educators and businesses are sounding the alarm that the traditional approach, which encourages passive absorption of facts and is teacher-directed, should become obsolete. How are educational institutions responding to the technological challenge? First, educational institutions must recognize and accommodate the following realities and trends:

1. The pace of technological change is so great that many college graduates will need to be involved in continuous education. Already, many companies are downsizing and requiring employees to acquire new skills.

2. The sophistication of today's communications infrastructure rapidly brings education to people in their homes and workplaces.

3. New educational enterprises are being developed to meet the educational needs of individuals, and if traditional institutions must meet their enrollment needs, they must become aggressive and be on the cutting edge.

4. Limited resources impede universities' capability to expand existing facilities to accommodate the increasing demand for higher education. Some states are becoming proactive in this regard and are engaging in initiatives that will accommodate the high demand for higher education.

Obviously, technological change is so ubiquitous that schools and colleges across the country and the world must embark on ways to integrate technology into teaching and learning. Besides the economic imperative, today's individuals are growing up in a technological age where everything around them is electronically constructed and applied. Children growing up in such an age will not respond appropriately to lecture-dominated learning. Thus, in light of the overwhelming force of the technological age, no educational institution can afford to ignore the technologizing of learning and teaching.

In this chapter, I will present one institution's attempt to encourage faculties to transform the way teaching and learning are conducted to meet present realities. Specifically, I will discuss the experience of facilitating an electronic modularized course for undergraduate students in a pre-service teacher-education program; the processes for designing and implementing an electronic modularized learning course; its strengths and challenges as perceived by students and faculty who participated in the project; and concluding suggestions regarding online modularization courses. The following questions will serve as focal points for the chapter: 1) Can a teacher education course be taught asynchronously and effectively to undergraduate pre-service students? 2) Can a multicultural education course be modularized? 3) Is there a qualitative difference between asynchronous and synchronous learning modalities relative to students' learning? 4) What are the strengths and challenges of an electronic modularized learning approach? 5) What are educators' responses to an electronic modularized course for undergraduate students?

BACKGROUND OF THE STUDY

In the spring of 1997, my institution received a grant from the state to encourage faculty to develop innovative approaches that would significantly improve students' learning and utilize the potential of information technology. Faculty had the understanding that the state was engaging in initiatives to accommodate the increasing demand for higher education in the not-too-distant future. The goal was for faculties to develop modularized courses that would utilize multimedia and other means of delivery so that "learners can enter the curriculum where they need to, move through it at their own pace, and actively pursue their learning at a time and place that suits them" (Northwest Missouri State University Center for the Application of Information Technology to Learning June 1997).

Fourteen proposals were approved, including mine. Faculty were supported with the equivalent of a summer, three-hour credit course and provided assistance through workshops and seminars that brought experts in the application of information technology to learning. Faculty were required to develop new courses or reconfigure existing courses into modules that would emphasize learning rather than teaching.

MOTIVATION FOR IMPLEMENTING ELECTRONIC MODULARIZED LEARNING

Several reasons launched me into participating in the modular learning project. 1) The concept and the challenge intrigued me. I wanted to be a part of the cutting-edge adventure by utilizing information technology to transform the way teaching and learning are conducted in my discipline. 2) There was monetary incentive. The university supported selected faculty at the equivalent of a summer, three-hour credit course. 3) I wanted to engage in action-research to determine the efficacy of an asynchronous course for pre-service students in a teacher-education program. 4) I wanted to know if a multicultural education designed to develop students' multicultural perspectives, reflection, attitudinal disposition, and to cultivate skills for designing pedagogical strategies that are culturally responsive, could be actually broken down and sequenced.

Ideally, a modularized course is best for knowledge that is discrete and sequenced. However, multicultural competency tends to be holistic rather than fragmented into discrete parts. In addition, most educators have strong concerns about an electronically modularized course for pre-service teachers. Because of the emphasis on experiential, social, and affective learning, teachers are worried that an electronically delivered course will not adequately prepare pre-service teachers to cultivate the knowledge, skills, and attitudes prerequisite to functioning effectively as classroom teachers. I, along with some other teachers, shared this concern.

Anyone who has taught a diversity course or multiculturalism in education course understands the tremendous benefits of class discussions, sharing of individual perspectives and experiences, simulations, and guest speakers who put personal touches to the issues. These activities promote students' critical reflection and perspective-gaining, which lead to a reexamination of assumptions, beliefs, and attitudes. Thus, I wanted to determine if I would achieve the same quality of learning experience in an asynchronous learning modality. Of the 14 faculty who participated in the project, I was the only teacher-education faculty who implemented modular learning.

WHAT IS MODULAR LEARNING?

Modular learning is not necessarily a new concept. It is more like the old independent learning or correspondence course with a different twist. Electronic modularized learning is self-directed learning that facilitates students' ability to move through technology modules at their own pace, with schedule flexibility. It essentially uses a variety of delivery systems including the Web, AltaVista, videos, assigned readings, simulations, and ongoing assessments, with the instructor serving primarily as a facilitator or resource for the course. Typically, the course is broken into units of study called technology modules, which are posted on the Web for students to complete. Because it is a self-paced course, students are not necessarily constrained by the traditional semester calendar. In other words, they could complete the course in as short or long a time as their individual needs and flexibility demand.

Designing the Electronic Modular Course: The Multiculturalism Course

I reconfigured the existing traditional multiculturalism course I taught into technology modules. I developed five modules, some of which had sub-modules (see Appendix A). The course was posted on my Web page for students to download. Some modules were short and some were longer. The modules were semi-sequentially formatted. An overview of the course was provided that described the course relative to goals, competencies, performance indicators, expectations, requirements, procedures, working assumptions, assessment and grading criteria. This overview is basically the course syllabus. Each module was organized as follows:

1. An overview of each module

2. Stated behavioral objectives to be accomplished

3. Materials/resources/packets and their location

4. Instructions for completing assignments and activities

5. Related concepts

6. Required assignments to be completed such as:

   • **Pre-instructional Activity**: Each module had a pre-instructional activity that was assigned specific points. This was more of an anticipatory set activity.

   • **Module Quiz**: Some modules contained quizzes for practice and were not graded. Students were strongly encouraged to go over the quizzes to learn pertinent information.

   • **Video Activity**: Each module had assigned related videos and accompanied exercises to be completed for grading.

   • **AltaVista Activity/Group Activity**: To facilitate students' opportunities for affective and social learning and interaction, students were divided into teams for group activities, such as simulation, debates through AltaVista, etc. Each module had an AltaVista activity. Although the course was designed to be self-directed and individualized, I wanted to build in some group activities that would enable students to come together to learn as a community and meet some social goals. A related module issue was posted and teams were required to take a pro or con stance and to develop arguments with data to support their stance. Each team posted their stance and arguments on AltaVista. The rest of the class read the positions and arguments and made their own responses. This was similar to the crossfire activity required in the traditional synchronized course, except that in the traditional course, students debated the issues face-to-face, with opportunities for feedback from the class. Each of the AltaVista activities had a deadline, and all students were expected to complete their responses by the due date. The AltaVista provided a valuable medium for social interaction and affective learning. Points were assigned for each participation. A group simulation was scheduled and each team was required to work together to complete the activity.

- **Web Sites Activity**: Each module had listed Web sites, some of which were assessed and assigned points.

  - *Module final assessment.* Each module had a performance-based assessment that required students to apply knowledge gathered from the readings, videos, AltaVista, text, Web sites, etc. Because multicultural competencies are not a set of facts to be memorized and regurgitated, assessment was mainly in the form of essay and scenario responses that required deep reflection and analysis of cultural and social conditions and their impact on teaching and learning in a diverse society. Students were expected to demonstrate success by writing and submitting quality reflective papers that reflected knowledge and skills related to the module and course (see Appendix B).

  - *Module summary requirement and point system.* Each module had a summary of the module requirement and points assigned to each required activity (see sample module/Appendix B).

METHODOLOGY

Two sections of the multiculturalism in education course were offered in the spring of 1998, one with the electronic modular learning approach and one with the traditional synchronized approach. Initially, a maximum of 20 students was allowed to enroll in the modular course. At the start of the class, I provided an overview of the course, including the expectations and procedures. I emphasized the nature and structure of the course as self-directed and self-paced, with the instructor serving primarily in the capacity of a resource person and a facilitator, providing assistance as requested and needed. Students were excited at the freedom and convenience of pursuing their learning at a time and place that suited them without attending class. However, students were cautioned about the discipline and the high level of involvement in the course. Five students dropped the course after understanding the demand and responsibility involved. To avoid any misunderstanding, students were required to sign a contract testifying to their understanding of the nature and structure of the course.

Students were required to attend a total of five class meetings over the semester. The first two class meetings were devoted to acquainting students with the course, including requirements, materials, and an introduction to the fundamental underpinnings of multicultural education. The third meeting was spent training them to use AltaVista for posting and reading documents. The fourth meeting brought students together for a cultural simulation called Bafa Alpha. A clinical experience was required in which the students were paired with a culturally different student from the public schools to engage in a cultural learning exchange. The final meeting brought the students together to share and learn about their clinical experience as well as their experience of the course relative to the strengths and challenges. A pre-test and post-test were administered to assess the knowledge gained.

STRENGTHS OF
ELECTRONIC MODULAR LEARNING

Active Learning Engagement

Students cannot be passive in the course. They must be active to learn and perform in the course. Because the students primarily direct the course, it forces them to engage and interact with the materials and, therefore, forges a cognitive response. There was ample evidence that the students were actively engaged with the materials. In a traditional class, however, students may be engaged in class activities, but it is unlikely that they would interact and process information related to the course and demonstrate substantive knowledge as this group did. The post-test showed a remarkable difference in terms of content knowledge. The average of students' performance on the pre-test was 20%. The post-test average for the few students who completed the test was 90%. The reflective papers required students to integrate materials from the various resources provided. Students referenced materials and examples in their papers, which demonstrated active engagement with the course materials.

Students constantly stopped by to discuss the issues one-on-one. Their reference to materials showed that they had interacted with the materials. This was very rare in the traditional class. Although not all students took the necessary responsibility, the quality of participation of students in the modular group was higher compared to students in the traditional course. Students did not have to take turns waiting to participate in the learning as in a traditional course; they participated at any time.

Collaborative Learning and the Learning Community

Contrary to expectation, this sense of community was an amazing development in the electronic modularized course. Being a self-directed course, one would have expected students to go about working individually. Contrarily, students sought out each other and formed learning communities. In a traditional course, when students are assigned group projects, they complain about working in groups. In the modular course, students voluntarily formed their groups to learn from each other. In one of the debriefing sessions, students commented that although the course was self-directed, they realized that the concepts were difficult to understand, but sharing ideas and discussing the materials in groups or pairs facilitated a better understanding of the content. The group activities such as the AltaVista and simulation enhanced the collaborative and cooperative learning.

Learning to Learn

One of the skills needed to survive in an information age is learning to learn. An electronic modularized course helps students to learn to manage their own learning. Although it demands a great sense of responsibility and self-discipline, it facilitates students' ability to acquire the skills to learn on their own. One of my colleagues commented that students did seem to get to a level of minimal competence more reliably than in a traditional non-modular course, and that the course forced them to take the initiative to keep themselves on track, thereby lowering the "absolutely clueless" quotient. Two students' comments are worth sharing here:

> I personally thought this course provided a very good experience for me. I have never taken a self-directed course before. It gave me the opportunity to acquire skills for independent learning.

> I learned to interact and process information and construct knowledge on my own. I had always relied on instructors to tell me what was important to learn. This time I was in charge of what I was learning and how I was learning it.

Self-Pacing

As a self-directed and self-paced learning course, students paced themselves accordingly, although too slowly to master the learning. Students completed and turned in modules according to their own pace.

INCREASED COMMUNICATION BETWEEN STUDENT AND PROFESSOR

I did not anticipate the increased and frequent contact with students that I experienced. I made the requirements and instructions so clear that I was expecting to receive students' completed modules in the mail. On the contrary, there was increased communication between the students and me. I had more contact with students in the modular learning course than students in the three other courses combined that I taught. Students constantly e-mailed, telephoned, or dropped by the office to ask questions, check on their papers, or discuss the materials they read. Between late January and early May when the semester ended, I documented close to 200 e-mails from the 14 students enrolled in the course.

On the positive side, this was a group of students I bonded with extremely well. When they dropped by to turn in or pick up their papers, they stayed to talk about the topics and issues. Multicultural issues are interesting though controversial. But the students and I talked about the issues in a conversational tone and in a non-threatening manner. Because students were free to express their thoughts about the issues, and because of the tact with which I stirred them to look at the issues from different perspectives, there was an increased positive effect toward the course and the professor.

CHALLENGES OF
ELECTRONIC MODULAR LEARNING

Student Responsibility

An electronic modular course is very demanding. Because the course is self-directed and self-spaced, a great sense of self-discipline is required. Many students who enrolled in the course did not have the self-discipline needed to complete the work. Procrastination was the primary exhibited behavior. Because of the lack of deadlines, most of the students tended to push the course aside with the intent of completing it tomorrow or the next day. Some of the modules in the multiculturalism course were very short, short enough that they could be completed in one day, yet some students did not turn in the first module, the shortest of all, until May 1998, and some have yet to turn in the first module (as of July 1998). By the deadline for dropping a semester course, two students came to me to say that they could not complete the course because they had procrastinated for too long. They added that the type of learning required in the course did not match their learning style. This was good information because educators need to be cautioned about rushing into new horizons without reasoned and informed knowledge.

Students' feedback showed that deadlines were important and necessary. But really, modular courses should not have deadlines as they are designed to be self-paced and self-directed. One of my colleagues noted that "although these are college students, the majority of them are not yet responsible learners." Generally, some students do not expect to work hard for their own grades. Modular learning forces students to work hard on their own just to become familiar with the content. Students' low expectations for academic work and associated behaviors overwhelmed some students' sense of responsibility and performance. One student's comment sheds some light: "I learned I needed to be more disciplined with time management. I realized how dependent I had been on deadlines and instructors pacing my course work."

Of the 14 students who officially remained in the course, only three completed the course in May when the semester ended, with two performing exceptionally well and one turning in poor work with the intent of accepting a "D" grade and superseding it through the traditional course. The most successful parts of the course were the AltaVista and cultural simulation requirements, which had deadlines. It was interesting to watch how students rushed to complete the AltaVista activity as the deadline approached.

Increased Faculty Workload

Designing and implementing an electronic modular course was time consuming. Designing online guidance and the modules, selecting resources and Web sites, assessing and grading essay papers, monitoring students' work, and managing students' correspondence were extremely overwhelming and time consuming. Even though the class did not meet regularly as in the traditional courses, the time devoted to the modular course exceeded the amount I expended on my other three classes combined. As with my other classes, I always allowed students opportunities to revise their work to attain quality. This practice compounded the increased workload. Because some students felt the pressure to submit a completed module, they would turn in work that was poor in quality, which required me to give lengthy comments and redirect them. Revising the work and resubmitting it to be re-graded was added workload and sometimes burdensome.

Student/Faculty Frustration

Some students expressed a degree of frustration with the modular course because of information overload and increased work, although most students complain about information overload and work regardless of the course's approach. However, I will agree that an electronic modular course, especially in a course like multiculturalism in education, can be beset with information overload and increased work. Because of the virtual nature of the course, the instructor is forced to present tremendous amounts of information and numerous resources and requirements to enable students to acquire pertinent knowledge as well as to ascertain that the students interact with the materials. The information overload and increased work overwhelm students.

Frustration for faculty also is an issue as some students seek easy ways to earn a grade without interacting with the materials. For example, students would want to complete the assignments without utilizing provided resources. Unfortunately, the assignments and assessments were constructed in a way that necessitated students' interacting with an array of information and resources to respond qualitatively. Some students, wanting an easy way out, would call to say that they were confused about an assignment and did not know how to complete it. But when asked to specifically identify what was unclear, they could not say, but would say "tell me what I am supposed to do." Again, when asked if they had read or viewed videos or visited Web sites related to the assignment, they had not. The worst frustration for the faculty comes from the backlog of work resulting from the delayed grades to be added onto the next semester's full-credit load. Administration's lack of sensitivity toward the faculty's needs could be debilitating.

STUDENTS' PERCEPTIONS OF THE ELECTRONIC MODULAR LEARNING COURSE AS COMPARED TO A TRADITIONAL COURSE

Students' perceptions of the electronic modular course as compared to a traditional course are summarized as follows:

Traditional Learning Course	Modular Learning Course
Teacher sets the pace for course.	Student sets and works at own pace.
Teacher acts as dispenser of knowledge.	Teacher acts only as a resource person and facilitator of learning.
Teacher controls learning by way of deadlines.	Student controls learning. Learns how to learn.
No room for procrastination.	High probability for procrastination.
Opportunities for class discussion, affective learning, and social interaction.	Limited to AltaVista.
Benefit of a classroom culture and close ties to course mates.	Lack of classroom culture.
Dependent learner.	Independent learner.
Passive absorption of information.	Active learning engagement.
Lack of self-discipline for learning.	Demands a great sense of self-discipline and responsibility. Prepares you for the "real world."
Use of guest speakers that enhance learning.	Lack of opportunities to hear guest speakers.
Lack of freedom to explore on your own.	Freedom to explore learning on your own.

Faculty's Perceptions of the Electronic Modular Learning

A survey was administered to the few faculty involved in piloting the project to determine their assessment of it. Although 14 faculty members had been approved for the modular project, only a few were able to develop and implement the technology modules. Some faculty did not modularize the whole course but experimented with one or two module components. Others were overwhelmed by the project and did not complete the design, let alone the implementation. Thus, only five faculty designed and implemented the modular course. Interestingly, their perceptions were similar to the analysis I provided in this chapter and to the students' perceptions relative to the strengths and challenges of modular learning.

On the positive side, faculty unanimously consented that if we are to prepare students for the twenty-first century, students need to develop skills for independent learning and for learning to interact, process, and apply information. Thus, a modular learning course helps them to take responsibility for their own learning and become active learners. The faculty's greatest concern was the lack of a sense of responsibility and self-discipline in students. They concluded that a majority of college students have been conditioned to be dependent on professors, "that only a select few can handle the freedom and responsibility that come with online modular learning." Some faculty felt that learning in some disciplines cannot be compartmentalized and that skills cannot be sequenced but must be acquired holistically. Only one of the faculty members felt that online modular learning is a disaster for undergraduate students. The other four felt that it could be a meaningful way of teaching and learning but that the associated challenges need to be overcome: work load; credit; and student self-discipline and responsibility.

Conclusions

What are my conclusions about online modular learning courses? Considering the way students have been conditioned to learn from elementary through high school, asynchronous learning networks do not work for all students. There are students who will continue to need the traditional structure. But this does not mean that these students cannot succeed. In the multiculturalism course, some students demonstrated that it could be done. However, they will have to adjust their learning style and their sense of responsibility. Thus, it is suggested that the pipeline of online learning with modularity be opened from the early grade years. If students get comfortable with using technology as a tool for learning and if teachers develop technology modules, students would gain the foundation and skills for a successful asynchronous learning network.

I believe that students can achieve quality learning if they take the responsibility. From this action research, I came to the conclusion that most students who do not learn well, do not because they have not learned the discipline of learning. They fail to take responsibility for their own learning. Modular learning forces students to take responsibility for their own learning.

Both technical and moral support from the administration is needed to achieve this success.

Contrary to expectations, it is possible to achieve social presence, and social and affective learning, if activities are carefully built into it. The AltaVista modality for interaction helps to accomplish this goal. The only drawback is that students may not be discrete about what they convey and post on AltaVista. There were instances where students were emotionally charged and posted some comments that were insensitive and naive. Based on my experience of the traditional course, with a face-to-face interaction, this type of intellectual maturity is unlikely to occur. Instructors implementing the online modular course may need to monitor such inappropriate online behavior.

Until we are able to deculturalize students' conventional way of learning, it is necessary to proceed with online course experimentation with caution. It is suggested that faculty, regardless of discipline, incorporate components of technology modules into their traditional courses to begin to gradually introduce students to self-directed and self-paced learning. Students who have experienced years of teacher-controlled learning will not automatically be comfortable with a learning approach that puts them in charge of their own learning.

The electronic modular learning project helped me to realize how heavily students depend on instructors to manage their learning. As we approach the twenty-first century, educators need to make a shift in paradigm that moves students from passive and dependent learners to learners who actively construct their own knowledge through utilizing the full potential of information technologies.

References

Jaeger, Werner (1945). *Paideia: The Ideals of Greek Culture*. New York: Oxford University.

Wirth, Arthur (1992). *Education and Work for the Year 2000: Choices We Face*. San Francisco: Jossey-Bass Publishers.

Northwest Missouri State University Center for the Application of Information Technology to Learning. Northwest Missouri State University, Maryville, Missouri.

APPENDIX A

Technology Modules for the
Multiculturalism in Education Course

Module 1:   Understanding Multicultural Education Concepts

Module 2:   Macrocultures and Microcultures

Module 3:   Understanding Multicultural Issues

Gender in Education

Race in Education

Class in Education

Linguistic Diversity in Education

Module 4:   Classroom for Diversity:
Rethinking Curriculum and Pedagogy

Module 5:   Clinical Experience:
Cross-Cultural Learning

APPENDIX B

Sample Module

Module 1: (1 week) Multiculturalism in Education:
                   Conceptual Understanding

Equality is the heart of the essence of democracy.

*America is not like a blanket—one piece of unbroken cloth, the same
color, the same texture, the same size. America is more of a quilt— many
pieces, many colors, many sizes, all woven and held together by a common thread.*

—Jesse Jackson

Overview:

This module explores the concept of multicultural education, the principles, goals, concepts, and assumptions that form the basis for responsive multicultural education programs. At the completion of the module, students
should be able to:

1. Explain multiculturalism in education and its fundamental purposes

2. Explain the goals, assumptions, concepts, principles, and issues of
   multiculturalism in education

3. Discuss the myths and misconceptions associated with multiculturalism

4. Explain the historical perspectives and legal precedents of multiculturalism

5. Discuss the controversial issues related to multiculturalism

6. Explain the need to reconstruct America's political, social, economic,
   and cultural life to reflect its diversity.

Materials & Resources

Readings: Textbook: Gollnick & Chinn; Chapter 1, *Multicultural Education
in a Pluralistic Society*, 2nd Edition, 1986, Charles E. Merrill, Publishing,
Columbus, Ohio,

Supplemental Readings: (see packet)

Handout 1a, No One Model America

Handout 1b, Parable of the Wheat Culture

Location of Resources: Owens Library and packet

Celebrating our Nation's Diversity

Our Diverse Nation

Celebrating our Nation's Diversity-Materials

Diversity Issues

Instructions

1. Complete the pre-instructional exercise 1a

2. Read handouts 1a and 1b (No One Model America and The Parable of the Wheat Culture)

3. Read chapter 1 of Gollnick & Chinn's book

4. Review module-related concepts

5. View video 1a (*America's Multicultural Heritage*) and complete the accompanying exercise

6. Participate in the AltaVista assignment

7. Complete module 1 assessment and submit to the instructor

Module-Related Concepts

Culture: A complex integrated system of beliefs, values, symbols, and behaviors shared by a human group that is dynamic. It includes a shared language, folklore, ideas, thinking patterns, communications styles, worldview, similar expectations of life, and the tangible artifacts/lifestyles associated with a people.

Macroculture: The national culture prefigured on the democratic ideals that gird the political and social institutions that is shared by most citizens.

Microculture: A subculture within the United States or a multiethnic society with unique cultural pattern and perspective not common to the macroculture.

Monoculture: The dominance of one culture over other cultures within a pluralistic society. It also means functioning from one cultural framework/worldview.

Cultural Assimilation: A one-way process in which persons of diverse ethnic and racial group give up their original culture and are absorbed into the core culture, which predominates in the society. The melting pot ideology or metaphor represents cultural assimilation. The idea of assimilation reflects the recognition of a superior culture and less superior culture. The idea of assimilation is antithetical to the democracy that is principled on the ideals of equality, justice, freedom, and opportunity for all.

Cultural Pluralism: A process of compromise characterized by mutual appreciation and respect between two or more ethnic groups within the context of a larger society. In a culturally pluralistic society, members of ethnic groups retain many of their ways of life, as a long as they conform to those practices deemed necessary for survival of the society as a whole. The salad bowl metaphor, along with stained-glass window, rainbow, tapestry and mosaic metaphors are associated with cultural pluralism. This is an ideal state of societal conditions characterized by equity and mutual respect among the existing cultural groups.

United States Citizenship: The 14th amendment defines citizens of the United States as all persons born or naturalized in the United States.

Multiculturalism: A philosophical position and movement that assumes that the gender, ethnic, racial, class, and cultural diversity of a pluralistic society should be reflected in all of the institutionalized structures of educational institutions, including the staff, the norms, the values, the curriculum, and the student body.

Multicultural Education: A reform movement designed to change the total educational environment so that students from diverse racial, ethnic, and gender groups; exceptional students; and students from each social-class group will experience equal educational opportunities in schools, colleges, and universities.

Diversity: The conceptual understanding that people are similar and different due to factors of race, ethnicity, gender, religion, exceptionality, class, lifestyles, etc.

E Pluribus Unum: Out of many, one. This refers to America's formation as a nation; that out of many nations, one nation, the United States was formed.

Pre-Instructional Activities

    a. Read handout 1a (No One Model America—see packet for module number one). Summarize the implication of the article for teaching and learning in a multicultural society.

    b. Read The Parable of the Wheat Culture (see packet for module number one): What is the parable telling you? How does the parable relate to contemporary United States society? What is your response?

Video Activity

View the video *America's Multicultural Heritage* (Owens Library). Complete the video exercise. Refer to rubric/guide when completing the exercise (rubric/guide in module packet).

AltaVista Activity

Posted; to be completed by 2/19.

Module 1: Final Assessment: A reflective paper

1. Develop a well-organized analysis with supportive data relative to the strengths and challenges of the following proposition: "The United States society and its institutions should promote cultural pluralism."

2. Design a construct that depicts the relationship between the United States macroculture and its microcultural groups. Explain and justify your design.

Module 1: Summary Requirements and Point System

1. Pre-instructional activity (15 points)

2. Video exercise (15 points)

3. AltaVista assignment (20 points)

4. Module 1 final assessment (50 points)

CHAPTER 23

Technology and Research in the Teaching of Foreign Languages, Literature, and Culture

Nancy Mandlove
Professor, Foreign Languages

Wofford College
Spartanburg, SC

BACKGROUND

In 1991, Wofford College completed construction of the high-tech Olin Building, which provided the initial resources for a new, technologically oriented curriculum in the department of foreign languages. The building is equipped with a teaching theater, a computer center, a 46-station Sony 9000 language laboratory (including 24 video stations), and a multi-band dish antenna dedicated to foreign languages. Each classroom contains a video presenter, videocassette recorder, television feeds, large projection screen, computer, network connection, and telephone links. Two major grants from the Mellon Foundation (to Wofford College and Furman University) have supported faculty training in the use of technology and extensive course development.

This year, a new grant from the Culpeper Foundation to the department of foreign languages will enable us to convert an existing seminar room into a multimedia center for faculty and student use. The center will include networked computers with multimedia capabilities: video and audio capture and editing; slide and flatbed scanners; color laser printers; and other hardware and software. The

Culpeper grant has also provided new Mac G3 computers for all faculty offices in the department of foreign languages, as well as the addition of new and upgraded computers in the language laboratory. These resources will allow the department to integrate the use of technology into all levels of the curriculum and will allow us to train student assistants to work with their peers in using and developing multimedia programs.

Incorporating Technology into the Curriculum

The acquisition of equipment and the incorporation of technology into the foreign language curriculum have been a long and gradual process, beginning in 1979 with two portable cassette players and advancing five years later to an 18-station, audio-only facility. Our current situation permits us to incorporate the use of technology at all levels of the curriculum and to require all language majors to become proficient in the use of computers, multimedia hardware and software, and presentation programs. Student involvement with technology falls into three basic categories: use of prepared materials (commercial materials or materials prepared by faculty); use of technology for class assignments and projects; and preparation of multimedia programs for use in the department by other students.

At the beginning and intermediate levels of language acquisition, students are required to use technology for class and lab assignments and to demonstrate basic skills on the computer: word processing; e-mail; Internet; and simple presentation program applications (HyperStudio, PowerPoint). In addition to the regular use of commercial materials (SCOLA-Satellite Communications for Learning, which transmits foreign language programming by subscription, videos, films, and CD-ROM), the foreign language faculty have prepared original materials on the computer for student use in class and lab. Examples of two recent projects for intermediate Spanish classes are a multimedia program on the Nicaraguan elections and one on the life and work of a Mexican folk singer who visited the campus.

The Specific Projects

For the first program, we acquired a videotape of the campaign advertisements of 24 candidates for the Nicaraguan presidency and integrated the ads into a multimedia computer program. The program was used in class and lab for aural comprehension, dictations, and discussion of culture and politics. Students were then required to follow the election campaigns and results on the Internet and to report their findings in class. For the second program, advanced Spanish students conducted videotaped interviews with visiting Mexican folk singer, Gabriela De la Paz. Video clips of the interviews and recordings of her songs were then incorporated into a multimedia computer program and used for listening comprehension exercises in second-year language classes.

Having acquired basic computer skills in the introductory language courses, students are prepared to undertake more sophisticated and original projects using technology in the 300- and 400-level literature and culture courses. Spanish 308, for example, is a course designed to introduce students to the study of literature and culture and to prepare them for what is expected in the major programs. One of the most important assignments in that course is the construction of a multimedia, hypertext program combining the skills needed for the analysis of literature and culture with the technological skills and research tools necessary for advanced work in the department of foreign languages. Each class is assigned a poem. Students then work in teams to produce an annotated and illustrated text based on that poem.

Last year's project focused on a poem by Pablo Neruda. Each team was assigned a particular area of research: the life of the poet; the poet's work; literary devices; the political climate of Chile; and Chilean culture. Each team then produced a multimedia project on that topic, integrating text, images, sound, and video. Individual writing assignments for that section of the course consisted of short explications of stanzas of the poem and examples of poetic devices used in Neruda's poetry. After considerable editing and correction, the explications were integrated into the program so that a student reading Neruda's text can click on a difficult passage and find one or more explanations or interpretations. Students can also find examples of the major poetic devices (metaphor, simile, synesthesia, etc.) drawn from Neruda's work. At the end of the semester, all the team contributions were integrated into one multimedia program that can now be used by students in future classes. Each year new poems, annotated and illustrated, will be added to the project.

In the Latin American and Caribbean Studies program, housed in the department of foreign languages, technology is central to the acquisition and presentation of information at every level of the program. The Internet is the single most important resource for current news of the region; contact with major university research programs; access to individual researchers throughout the world; and up-to-date information on geography, politics, culture, and statistics. In the two introductory seminars, student teams present daily news updates on the region with text and images gathered from the Internet. Each team also produces two multimedia projects each semester using video, audio, Internet resources, images, and text. The projects are designed to integrate course content and to foster skills in research, presentation techniques, and teamwork. All students in these seminars spend three hours per week in independent laboratory sessions using video materials, computer simulations, and Internet resources.

Students in the Latin American and Caribbean Studies program are required to complete a senior capstone course in which each student produces a multimedia project on CD-ROM, which then becomes a permanent part of departmental laboratory and classroom resources. The Wofford College foreign language department combines emphasis on technology with a strong study-abroad program. Students in the Latin American and Caribbean Studies program are encouraged to identify an area of research for the capstone project prior to study abroad. In addition to collecting resources and materials, they can also take their own photographs and produce their own video materials during their foreign study experience.

The area studies program is relatively new at Wofford. In 1996, the first students completed the program. Those students were required to submit a traditional research paper for the capstone course. Students completing the program in 1997 and 1998 (11 in total) were subject to the new requirement of completing a multimedia computer project. Some of those have been excellent, combining serious original research with effective use of technology.

One student produced a CD-ROM on the topic of tourism and ecology in Costa Rica. It included original videotaped interviews with former president and Nobel Peace Prize winner Oscar Arias, the Minister of Tourism, Costa Rica's ambassador to England, superintendents of several national parks, and footage demonstrating the benefits of tourism as well as its detrimental effects on the environment. Other projects have included a study of Argentina's Dirty War, grassroots organizations in Guatemala, and immigration—including original videotaped interviews with Mexican workers who participated in Wofford's student volunteer ESL (English as a Second Language) program.

Of the 11 students who have completed the multimedia project requirement, four (36%) have returned to campus *after* graduation to continue working on the project. (Needless to say, none of the students who completed the traditional research paper returned for further work on the paper, nor has any other student done so in my 30 years of teaching!)

THE EFFECT OF TECHNOLOGY ON STUDENTS

The most exciting part of these curricular changes is that students take much more responsibility for their own learning, and they work together as teams to produce something that will be seen and used by other students. The Internet has revolutionized the teaching of foreign languages by providing immediate access to news, people, cultural information, radio broadcasts, and images from all over the world. It allows students to find and contribute valuable information to the classroom in a way not previously possible. Students feel a sense of ownership and a direct connection between their own experience and their academic work, particularly when they use their study-abroad programs as the springboard for subsequent class projects.

Learning becomes much more student-centered, while the teacher becomes a kind of "orchestra leader" who helps to integrate and assess information that the students, themselves, contribute to the class. The traditional "research paper," while valuable for teaching certain skills, ends up in a file—never to be seen again. By contrast, students know their multimedia projects will remain a part of the departmental curriculum and are motivated to produce an interesting and meaningful piece of work.

Students recognize that the skills they acquire in collecting, editing, and presenting information through technology will serve them well as they enter the job market. They also recognize that the ability to work as a team member is an essential part of the current environment in business, government, and other fields. Teamwork and the laboratory settings create a sense of community and camaraderie among students in the department, which is quite different from the isolation of traditional scholarship and contributes to student morale. Even the inevitable glitches, "crashes," incompatibilities, and disasters that are a part of the world of technology contribute to a sense of shared experience.

DIFFICULTIES ENCOUNTERED

While we are firmly committed to the incorporation of technology at all levels of our program, it is not always easy to succeed. At the most elementary level, when using the Internet as a research tool, it is necessary to teach students to evaluate the source of the material and to differentiate between "Joe's Home Page" and *The New York Times*. At the upper levels, when students begin to create their own multimedia presentations, the relationship between content and presentation or substance and creativity, needs to be addressed. Some students enjoy experimenting with the technology at the expense of substantive content; others produce what is essentially a term paper on PowerPoint slides, with lots of text and little integration of other relevant media. Since most faculty have been trained in traditional scholarship, it is sometimes difficult to help students achieve the balance needed to use electronic means of communication effectively.

Hardware and Software

The most persistent (and annoying) problem, however, is the technology itself—hardware and software problems. Large numbers of students working independently on multimedia projects, using the general college computer labs or their own home computers, will encounter incompatibilities and numerous other malfunctions. It is important to provide very specific written guidelines regarding the use of machines and software. Unfortunately, the guidelines usually arrive as the result of someone's spectacular failure.

There are many multimedia software programs available. It is important to weigh the advantages and disadvantages of various programs in terms of what they offer compared to the amount of time it takes to teach students to use the program effectively. Since the goal is to use technology to gather and disseminate substantive content, the time devoted to learning technical skills needs to be minimized. While relatively simple applications like HyperStudio and PowerPoint offer less flexibility and sophistication than a program like Director, students can learn to use them well in a very short time.

INCREASED INTEREST

While it is not possible to say with absolute certainty that emphasis on new technology is responsible for the dramatic increase in foreign language enrollments we have experienced at Wofford, the numbers are impressive and we believe that they are largely due to the facilities we have in the high-tech Olin Building. Overall student enrollment at Wofford has remained steady over the last 20 years at approximately 1,100. In 1979 (with a language laboratory consisting of two portable cassette recorders), the department graduated five language majors. Over the last five years (since the completion of the Olin Building and the rapid acquisition of new technology), the department has graduated 115 language majors—an average of 23 each year. In May of 1998, we graduated 37 majors.

Ten percent of the Wofford student body now majors in foreign languages and the department has grown from four full-time faculty and two part-time to six full-time and two part-time. The retention rate in foreign language courses, from required courses to voluntary, is approximately 40%. During the fall semester of any given year, 45% of all Wofford students use the foreign language laboratory weekly. The Latin American and Caribbean Studies program, which began with an initial enrollment of six students in 1994, had an enrollment of 21 for the fall of 1998. The first group to graduate having completed the 20-hour area studies certificate consisted of three students in 1996. In 1998, seven students completed the program; next year, 13 will graduate with the certificate.

CONCLUSION

We believe that teaching with technology has contributed greatly to the increased interest in foreign language study at Wofford. It gives students access to a wealth of authentic foreign language materials and helps bring the real world into the classroom. The extensive use of technology also brings students together in group settings (language, computer, and multimedia labs) where they experience a common sense of purpose. Students work closely with faculty in these informal settings, which boosts morale and motivation on the part of both students and faculty. Students have more control over their own learning and accept more responsibility for classroom activities. It is a whole new way of teaching and learning, and we have only just touched on the potential that technology holds for education.

CHAPTER 24

Integration of Instructional Approaches Through Media Combination in an Undergraduate Information Systems Course

Katia Passerini
Instructional Design Specialist

Mary J. Granger
Associate Professor of Information Systems

George Washington University
Washington, DC

INTRODUCTION

Research in educational technology shows that students' learning is facilitated by instructional strategies that take into account the different learning styles and abilities of the recipients of instruction. Providing instruction by using several different media types conveys the benefits each medium has on learning (raising attention, visualizing the topic, providing a certain level of class interaction, etc.). Integrating different media into single multimedia presentations for class use allows joint exploitation of these advantages as well as integration of several instructional strategies.

In this chapter, instructional approaches based on the use of different media are reviewed to provide examples of strategies and evaluation tools for media integration. An introductory information systems undergraduate course taught at the George Washington University is used as an example.

Description of the Course

The undergraduate information systems applications course is designed to introduce students to the technology and trends of business information systems. Students are exposed to topics such as telecommunications (including Internet and electronic mail), multimedia, productivity software, database management systems, and other topics in management information systems (MIS). The objective of this course is to help students integrate information systems and technology into organizational settings. Lectures are delivered as multimedia presentations in Microsoft (MS) PowerPoint '97. The course includes a hands-on laboratory section in which students work with integrated applications software packages (such as Microsoft Office).

The Students

The majority of the student population in this course is composed of sophomores in business administration. Their age range usually varies from 19 to 22 years old. But the age range increases if there are transfer students. Generally, this student population lacks prior work experience. A survey conducted in spring 1996, in four sections of this course, showed that the majority of the students used computers primarily at school. Work experience is limited to short internships and does not necessarily involve computer training and usage. The survey collected the demographic characteristics of the students entering the course each semester to refine the objectives and the depth of coverage of the subject material of the hands-on laboratory sessions. One of the assumptions of the data collection was that the new generations of undergraduates enter this introductory course with an increased knowledge of computer applications and, therefore, they do not need to be instructed in the basics. However, this survey reinforced the need to conduct the laboratory portion of the course at an introductory level.

The Course

Students are expected to attend their first business course as a seven-week module integrated with semester-long courses. The lectures are distributed over half a semester, although the time of each lecture session is two hours and thirty minutes, with an additional two hours for laboratory instruction on another day of the week.

In the laboratory part of the course, students work with graduate teaching assistants that lead their exploration of the applications software packages.

Group activities include team presentations, using Microsoft PowerPoint, of comparative analyses of the use of information systems by national and international companies in selected industries. Students focus on integrating text, sounds, and graphics to convey information about the industry. The goal of this project is to have students conduct research and explain the importance of information systems in the management of companies in selected industries and countries. Students conduct a financial analysis of the stock price fluctuations of the chosen companies and present key technology issues that are promoting (or inhibiting) these companies' competitive advantage in their sector.

The topics covered in this seven-week course are:

- History of computers
- Hardware/software
- Telecommunications
- Productivity tools
- Data management
- Multimedia

These topics are introduced through a blending of technologically-based presentations and in-class discussions, with the objective of reaching a working compromise between exposing students to using computers and fostering their thinking and problem-solving skills. Both of these skills are needed in today's job market and it is necessary to focus on them jointly. The laboratory component of this course reinforces this objective by offering hands-on experience.

To foster in-class peer learning, in-class review questions are included in the form of short quizzes. Additionally, in order to foster student retention and learning, weekly homework is required and includes answering short questions or elaborating on critical thinking questions.

THE SOFTWARE

Lectures are developed in MS PowerPoint '97. The reasons for the selection of this presentation software are:

1. Versatility of the software—The latest version of this application enables dynamic multimedia presentations with easy-to-use animation. The class illustration exposes students to the features of the software, which they are required to use in their final group project.

2. Ease of use of several channels—MS PowerPoint allows media integration without the need to author scripts. It enables the use of several channels of communication by instructors without programming backgrounds. It fosters the application of the principles of *communication theory* by promoting more than one channel to communicate with students.

In spite of the capabilities mentioned above, Microsoft PowerPoint is not a real multimedia application since it lacks the tools for guaranteeing user control of the events timing. (Stop, play, and pause buttons are crucial for using video in a learning environment.) However, it is an acceptable substitute that does not require an extensive time investment in authoring, it is available with Microsoft Office, and is compatible with Windows 95 installed on Pentiums.

THE FACILITIES

The course, although currently taught in typical classrooms, is planned to be offered in a multimedia classroom equipped with Pentium computers. The computer set-up is ergonomically designed for online and off-line instruction. The setting of this classroom facilitates *peer learning* and interaction, as the monitor, motherboard, and keyboard are built in to students' desks, which remain regular note-taking tables when the computer is not in use. This new computer distribution facilitates communication and interaction among students. A computer monitor located between the student and the instructor (the usual lab configuration) or between the instructor and the classroom creates a fictitious distance between learners, and lessens efforts toward cooperation. All computers are connected to a LAN, and communication can now occur directly or indirectly through chat rooms or "talk" functions.

The instructor has access to several teaching aids: television; videoconference equipment; overhead projector; and blackboard. The design of the developed presentations requires a LCD projector. Video, sound, text, and animation are all included in the slides for the lecture. The intent is to use "multimedia" rather than multiple media to convey the message.

THE INSTRUCTIONAL APPROACHES

The literature on multimedia indicates that the use of the integrated media approach facilitates more interaction and user control of the application than individual media. *Multimedia* combines several media such as text, sound, graphics, video, and animation for simultaneous delivery through a computer or other electronic means (Vaughan 1996). Reading a textbook and watching a content-related video is not the same as using multimedia. It is using multiple, *separate* media rather than a set of synchronized tools. *Interactivity* adds another component to the definition, which is the user control of the events delivered by the computer. The user is required either to initiate, pause, or stop an action or to provide feedback to demonstrate that learning has occurred (for example, answering questions that are asked at the end of each session).

There are several objectives behind the use of multimedia *in lieu* of multiple media. The most compelling arguments are that a multimedia title raises attention, motivates the user, generates different simultaneous interactions, such as feedback and performance assessment, and, thus, enhances retention and assimilation of information by providing "plenty of action and novelty" (Stemler 1997). The stimulating effects of combined media generate synergy (the whole is bigger than the sum of individual pieces) in a framework where user participation is constantly required via navigation tools and user-friendly interfaces.

Multimedia also combines the individual media in content, spatial, and temporal synchronization, thereby associating text, video, sound, and graphics in a homogeneous product that appeals to different users and gives them control over the sequence of learning events through the use of control buttons. The three categories of synchronization described in the software development and implementation section of this chapter (content, spatial, and temporal synchronization) need to be applied constantly for the creation of effective and aesthetically attractive multimedia.

By drawing upon the differing learning abilities of the students, the different tools can reinforce the instructor's explanations. *Visually oriented learners* benefit from animation and video presentations. Their ability to learn is best fostered by the use of motion. The users, via control bars, pace through a presentation that includes video and animation. *Linear-thinking learners* benefit from reading text sequentially. A detailed lesson summary, provided on the instructor's Web site, facilitates their interaction with the material presented.

Students needing sequential events to facilitate learning will even benefit from a textbook that does not carry complex graphics, colors, and references to many other pages or URLs. Some of them complain that using text presentation loaded with graphics shifts their focus and they prefer straightforward textual presentations rather than overcrowded graphical images. *Associative thinkers* benefit from a presentation that allows hyper-linking instead of page-by-page text reading.

In addition to the informative function accomplished through the above processes, the presentations developed for the course are designed to apply different learning theories. The learning theory represents the philosophical underpinning of, and offers guidance for, the content and design of instructional software. Choosing a specific learning theory provides different views on the relationships between the acquisition, representation, organization, maintenance, and transfer of knowledge. In each learning theory, the inputs, the products, and the processes of learning differ. The cognitive model identified by the theory will differ accordingly. For example, if it is believed that learning occurs as a response to a stimulus (behaviorist theory), the main concern will be the creation of the appropriate stimulus (the input of instruction). Or, if it is believed that learning is a result of the physiological development of the individual (developmental theory), the main concern will be the process and phases of learning rather than the objects of instruction.

In this course development, the authors follow the constructivist perspective (CP), which looks at learning as a process in which individuals construct knowledge by actively collaborating with others and conducting group activities. In order to stimulate this type of learning, students are exposed to business problems involving the optimization of the use of computers in organizations. They are asked to work in groups to offer examples for better management of information systems to solve the specific problems. Class exercises and games promote interaction and facilitate retention through practice and feedback.

Feedback may be delivered in the form of text messages (providing correct and incorrect answers or general error signals), sound clues (earcons with positive or negative reinforcement), flashing lights, or changing object colors. It may be all of the above, or it may be differentiated on the basis of the audience. The timing of feedback (instantaneous versus delayed) needs to be considered as well as the format (text versus graphic) and content of the feedback. A wrong answer may be associated with a link that connects to a review page, it may offer the option to retry, or it may provide other contextual help mechanisms. To foster reflection, feedback may be provided only at the end of a lesson test, game, or simulation. Interruptions before lesson completion may cause distraction and may decrease the reflection moment.

The instructional method used in the undergraduate course to increase retention is based on the synchronous in-class provision of feedback. The multimedia presentations contain true-or-false questionnaires that are interactively answered in class. Immediate feedback is provided by sound clues that are associated with the "right" and "wrong" answers. Critical thinking questions are also included in the presentation and students are required to thoroughly justify their answers both individually and in assigned groups. The instructor acts mainly as a discussion moderator.

INSTRUCTIONAL DESIGN

Once the focus, the underlying learning theory and the objectives of the course are identified, the *instructional methodologies* and designs to be used to reach those objectives can be matched with the theory. The instructional design mode reflects the underlying theory; for example, it may focus on programmed step-by-step instruction in a behaviorist model (Gagné 1985) or on cognitive readiness in a developmental model (Piaget 1977). The type of presentations to be designed vary accordingly, from a simple display of information to simulation and games, from teacher-centered instruction (or computer-centered, in this case) to user-centered learning.

In this introductory course, the design of the instructor presentations is based on the behaviorist principles identified by Gagné (1985), although constructivist components are used as part of the discussion-question sections of the presentations. There are nine objectives (see Table 24.1 on page 314) that a multimedia developer has to consider when producing educational multimedia modules.

To reach the above objectives, each element of the class presentations complies with definite guidelines. Each screen must provide effective instruction, appropriate navigation tools, and visual aesthetics (Milheim and Lavix 1992). In terms of the instructional objective and amount of text information, class presentations follow McFarland's (1995) findings that the amount of text included in a page depends on the age and grade level of the learners.

The presentations also incorporate other principles of multimedia development, such as interaction with the learners in the form of *stimulus-response-reinforcement*. Schwier and Misanchuk (1993) see this inclusion as facilitating elicitation and assessment processes. In terms of navigation principles, Gurak (1992) states that users recognize icons when they are associated with familiar symbols.

Table 24.1

GAGNÉ: INSTRUCTIONAL EVENTS & CORRESPONDING COGNITIVE PROCESSES

Instructional Event	Cognitive Process	Action	Media and Strategy Used
Gaining attention	Reception	Show the user an attraction event	Text & graphic animation, use of sound clips in titles
Informing the learner of the lesson objectives & activating motivation	Expectancy	Pose a question/make a statement that will identify what the learner will do & will be able to answer at the end of the lesson	Text & graphic, slide transitions, flying bullet points that identify lesson objectives
Simulating recall of prior learning	Retrieval	Review existing knowledge of topics relevant to the lesson. Ask questions.	Text, graphics, & video to solicit recall of prior topics
Presenting the stimulus material	Selective Perception	Introduce the content material	Text, graphics, video & sound
Providing learning guidance	Semantic Encoding	Show an example on how to create an object relevant to the lesson material	Graphics, video & sound with meaningful examples & instructions on how to do
Eliciting performance	Responding	Ask the student to provide examples	Text mainly, use of questions
Providing feedback	Reinforcement	Check whether examples are correct/incorrect & explain	Text, sound, and graphics with icons indicating right and wrong answers
Assessing performance	Retrieval	Provide scores of tests & remediation	Text, graphics, & sound, with the number of successful results & additional explanations for difficult questions
Enhancing retention & learning transfer	Generalization	Show other examples & ask students to draw parallels	Graphics, video, with focus on concept mapping

Adapted from Learning Theories,"http://www.gwu.edu/~tip/gagne.html,"06/20/1998.

Additionally, McFarland (1995) suggests extensive use of traditional pause, stop, and play buttons. There is no magic number of lines of text to be used and, therefore, the instructor needs to experiment and evaluate students' reactions.

Text and graphic animation is included in the design, as it contributes to attention gathering, although no extra learning effects are attributable to the use of animation (Hannafin and Rieber 1989). Use of audio is limited at the level of earcons, such as abstract sounds that display a status or feeling (in this case "right" or "wrong" answers). Wright (1993) believes that the combination of visual presentation with audio explanation delivers information in an easily understood format.

The videos selected in the class presentations are synchronized with the content and reinforce the concepts presented through everyday-life examples. Taylor (1992) suggests that, although video is not the most appropriate medium for presenting detailed material, it does have an emotional appeal that entertains and captures learners' interest.

Given the different set of learning benefits attached to individual media, the presentations' final interface is a multimedia application. The reason for choosing multimedia is related to this synergistic effect and the ability of these applications to capture audiences with differing learning styles. These audiences may be more at ease in learning with a video rather than a textbook and the presentations are intended to maximize the in-class learning. The lectures combine the use of several media to complement text presentations. Each presentation is organized by topic and each topic contains short videos, hyperlinks, and detailed graphics of the specified topic.

DEVELOPMENT AND IMPLEMENTATION

Effective screen design needs to include the coordination of text and graphics to present a content in a sequence that facilitates understanding. The visual clues on the screen cannot be cluttered: Too much representational text or information (declarative text) in one screen creates confusion and appears overwhelming. The use of text needs to be synchronized with the type of graphic on the same screen. The graphic content must reinforce the text. The timing must allow for a reading speed from the computer screen that is 28% lower than the regular reading rate (Stemler 1997).

The layout of information on the screen (spatial synchronization) should comply with the functional areas identified by multimedia researchers with the "navigation toolbox" at the bottom of the screen, "title and instruction" area on the top, and "body area" in the center (Stemler 1997). This example, developed for use of text, applies across different media. Sound clues can be used as *attractors* to specific areas or functions on the screen (use of earcons) or as *aloud reading* of material presented in text format (declaration) or as emphasis (representational) for specific events, with a changing tone of voice that reinforces the message. It is important that there is coordination across the media used for each function.

Content synchronization must occur in all media levels. If there is a character standing near the grave of a relative in a funeral scene, the sound effect must comply with the content (somebody crying and not laughing), so must the animation (slow speed of cars proceeding towards the cemetery entrance, not a car race) and the video. *Temporal synchronization* is crucial, as the different events in the screen need to start either on user control or in an uncluttered sequence that does not create media competition for catching the attention. *Spatial synchronization* deals with the distribution of the different media on the screen. This should preferably be allocated to the body area of the screen, leaving the side areas consistent with the navigational tools used in the whole application. As long as the different elements are synchronized at all levels, the instructor's presentation is highly effective.

The variety of media supports learning processes that rely on knowledge existing in memory and organizes it to produce new knowledge. In this delivery format, the instructor's oral presentation is supplemented by textual information and by the use of video, audio, and hyper-linking. Hyper-linking allows navigation beyond the current page or application. The presentation is enriched through links to other software applications that reinforce the explanation. For example, while discussing database management systems and presenting graphics of different types of relational databases, a hyperlink to Microsoft Access allows quick access to a different application that demonstrates the targeted principles.

Techniques for preparing class presentations follow the steps of multimedia development. They can be summarized as the following:

1. Content research

2. Development of storyboards

3. Assessment of clips needed (video, audio, and graphics)

4. Clips digitization (scanning, voice recording, video digitization)

5. Layout design and implementation (navigation icons)

6. Layout design and implementation (individual topic pages)

7. Testing and evaluation

8. Distribution on disk (packaging)

Step 1. Content Research

This preliminary stage requires thorough research on the information to be displayed in the application. In the case of multimedia, this implies going beyond the textual information. Pictures, photographs, and sounds are needed to make the application a true multimedia product. Since the purpose of the application is educational, a large percentage of the material provided in the textbook (which varies according to the type of media) can be used for the presentation.

Step 2. Development of Storyboards

Storyboarding is one of the most pivotal stages of the development process. It deals with both the *content* and *artistic* components of the prototype. It allows putting both primary and secondary text into the template layout, thus bringing immediate identification of space limitations. It also allows media organization, as each element on the template can be identified with the corresponding filename. The storyboard template is the most effective tool for drafting the design and having a visual representation of the prototype.

Therefore, a very careful selection of content materials needs to be made and adapted to the presentation format and screen layout. A lengthy page of text may become an attention detractor. It is necessary to find a balance between the information to be delivered and the continuous attempt to capture attention. When necessary, the use of text should be limited to bullet points. As an alternative to lengthy presentations, extended weekly summaries are available on the Internet, relieving students from notetaking and reinforcing class participation.

Step 3. Assessment of Clips Needed (Video, Audio, and Graphics)

This phase is an extension of the storyboard phase because the actual development of the storyboard depends on the type and amount of information available. The assessment of the material already available in digitized format (or eventually available after the appropriate digitization technique) is an indicator of the feasibility of implementation. In this phase, unrealistic expectations can be dropped and replaced.

Step 4. Clips Digitization
(Scanning, Voice Recording, Video Digitization)

This is the stage where several different techniques are used in order to obtain a file in a format (*.wav, *.avi, *.bmp) ready to be inserted in the application. Pictures can be scanned from the textbook with a regular HP scanner-jet. Voice is recorded with the default Windows '95 voice recorder and edited with Voice Edit software. Videos can be digitized using software such as Adobe Premiere and Asymetrix Digital Video Producer.

Step 5. Layout Design and Implementation
(Navigation Icons)

In this step, the basic principles of multimedia need to be taken into account, such as consistency of the screen design throughout the application and positioning of navigation icons. This consistency can be accomplished in PowerPoint by placing the navigation icons in the Slide Master.

Step 6. Layout Design and Implementation
(Individual Topic Pages)

For the development of individual topic pages, additions to the Slide Master default navigation icons include buttons that run video files (*.avi format). Ideally, in this stage the "BACK" and "NEXT" buttons are complemented by other buttons linking directly to a specific page (rather than just "NEXT") and back to the originating menu. However, because of the nature of PowerPoint, software navigation occurs sequentially. The sequencing is obtained by "mouse clicking" from slide to slide. Hyper-link navigation that allows non-sequential movement and permits navigation to different applications is user-controlled and occurs on user button clicks within the same slide.

Step 7. Testing and Evaluation

In this phase, the presentation functionalities are tested. Testing occurs both at the content and the technical level. Developers test technical integration, while potential users test content integration. Changes can be extensive or minor, depending on how effectively previous steps were completed and whether a formative evaluation was conducted while the presentation was being developed.

At this final stage, a summative evaluation is conducted on the performance, design, and functionalities of the overall application.

Step 8. Distribution on Disk (Packaging)

The last but not least important step requires packaging of the final product into a floppy disk (or zip disk). PowerPoint includes a feature that facilitates portability. The "pack and go" functionality lets the user save the whole presentation together with all the media clips that were embedded or linked in the application. It also specifies whether a PowerPoint Viewer is included in the packaged software for anyone who does not have Microsoft Office '97.

TESTING AND EVALUATION OF THE COURSE

The testing of the integration of the course takes place at the end of each semester. Evaluation forms are made available to students. To facilitate these evaluations, online surveys are administered. Anonymity is guaranteed by the use of networked computers in the School of Business's LAN, identifiable only by IP (Internet Protocol) addresses.

Another evaluation tool to be applied in the near future is the use of "external evaluators." Students' projects and papers will be distributed anonymously. The reviewers will not know which papers and projects are from students in the traditional classroom and which are from students in the technology-aided course. Once a significant number of papers and project are reviewed, the instructor will re-associate them with the originating classes and compare the results as reported by these external evaluators. The instructor will not participate in the evaluation to avoid "researcher bias." Reviewers will look at the variety of projects conducted throughout the course. For instance, in-class learning, assessed through weekly students' logs, will be evaluated by reviewers to gather understanding of students' development and assimilation of reading material.

In addition to the reviewers' input, the final exam tests the level of understanding of the subject as well as the effectiveness of the media used. It is developed in three parts, each favoring a different component of the instruction. One part will test the textbook material, another part will test class presentation material, and another the laboratory instruction. The scores in each section of the test will be compared to assess individual learning in different settings.

CONCLUSION

Business school faculty members have been putting more effort into integrating computers in the curriculum. Graduate courses mandate knowledge of productivity software as a pre-requisite for all courses. Workshops offered at the beginning of graduate programs have been designed to achieve standardization of computer knowledge prior to entering the degree program. However, most of the programs that rely extensively on the use of computers during the semester are related to technical fields, such as the Masters of Information Systems Technology, the Masters of Finance and the Masters of Accounting. Technology is used to a lesser extent in marketing, management, public administration, and strategic management courses.

The same unbalanced use of technology exists between IS (Information Systems)-related courses and business courses in the undergraduate curriculum. The introductory information systems course, which is mandatory for all undergraduate students, offers a perfect opportunity for a preliminary integration of computers into the classroom. Using this course in the format described provides an introduction to business information systems both at a theoretical level and at a practical level. It is expected that this teaching format will further students' learning by stimulating different learning channels. The use of several media is intended to target individual abilities to learn from different media (as per Gardner 1993, multiple intelligence theory) and, thus, to promote understanding of the subject matter by taking into account the needs of different learners.

Strategies for integrating the use of computers into the curriculum should look at the benefits of technology, its impact on the learning objectives, and the availability of alternative instructional methods. Additionally, an understanding of the different products of instruction enables the instructor to determine whether the subject matter is more effectively delivered through a tutorial, simulation, or games. Ideally, by taking full advantage of different media synchronized in one application, multimedia applications tailored to class content enable instructors to deliver the information more efficiently.

A LOOK AT THE FUTURE

Eventually this course can move into distance learning and the presentations can become stand-alone (offered through Real-time video). CDs on class content can be distributed for self-paced learning and class interaction can take place through list-servs, chat rooms, and Web communications packages such as Netforum.

An instructional technology laboratory, recently established at George Washington University, offers a unique opportunity to help faculty develop skills and strategies for moving to Web-based instruction, distance learning, and the integration of computers into the classroom. The beta-version of their Web software, Prometheus, is a step toward the consistent introduction, throughout the curriculum, of Web-based instruction in the classroom. The delivery of presentations can be automated and class discussion and exchanges can be promoted.

One section of this course used this software successfully. Although current students indicated that they do not have enough self-discipline to take the entire course online, they are eager to automate some aspects of the lectures through short-video and multimedia on Prometheus, while devoting more class time to group activities and panel discussions.

The introduction of computers in the classroom can increase the level of interaction among students rather than decrease it. The instructor and the students jointly share the responsibility for learning, and both can take full advantage of the integration of asynchronous processes with synchronous ones. The recipe for success matches the instructional methodology with the content of instruction, reorganizing courses as was done in this introductory course.

References

Gagné R. *The Conditions of Learning and Theory of Instruction.* New York, NY: Holt, Rienhart and Winston; 1985.

Gardner H. *Multiple Intelligences: The Theory in Practice.* New York, NY: BasicBooks; 1993.

Gurak L. Toward Consistency in Visual Information: Standardized Icons Based on Task. *Technical Communication.* First Quarter 1992; 33-37.

Hanclosky W. *Principles of Media Development.* White Plains, NY: Knowledge Industry Publications; 1995.

Hannafin M., L.P. Rieber. Psychological Foundations of Instructional Design for Emerging Computer-Based Instructional Technologies: Part II. *Educational Technology Research and Development.* 1989;37(2):102-114.

McFarland R. Ten Design Points for the Human Interface to Instructional Multimedia. *T.H.E. Journal.* February 1995; 67-69.

Milheim C., C. Lavix. Screen Design for Computer-based Training and Interactive Video: Practical Suggestions and Overall Guidelines. *Performance & Instruction*. 1992;31(5):13-21.

Newby T., D. Stephich. *Instructional Technology for Teaching and Learning: Designing Instruction, Integrating Computers, and Using Media*. Englewood Cliffs, NJ: Prentice-Hall; 1996.

Piaget J., et al. *Epistemology and Psychology of Functions*. Dordrecht; Boston, MA: D. Reidel Publishing Company; 1977.

Schwier R., E.R. Misanchuk. *Interactive Multimedia Instruction*. Englewood Cliffs, NJ: Education Technology Publication; 1993.

Stemler L.K. Educational Characteristics of Multimedia: A Literature Review. *Journal of Educational Multimedia and Hypermedia*. 1997;6(3/4):339-359.

Taylor C. Choosing a Display Format for Instructional Multimedia: Two Screens Vs. One. *Multimedia Display Formats-AECT '92*. 1992;768-784.

Vaughan, T (1995) *Multimedia: Making It Work*, (2nd edition). Osborne-McGraw Hill. ISBN 0-07-882035-9.

Wright, E.E. (1993). Making the multimedia decision: strategies for success. *Journal of Instructional Delivery Systems*, Winter, 15-22.

CHAPTER 25

The Electronic Esquisse: A Pentium Approach to 1700s Design Methodology

Antonio Serrato-Combe

Professor

University of Utah

Salt Lake City, UT

J'ai tout mon tableau dans la tête, mais je ne l'ai pas encore esquisse. I have everything in my painting in my head, but I don't have an esquisse of it yet.

5th Edition of the 1798 *Dictionary of the French Academy*

INTRODUCTION–HISTORY

In the 5th edition of the 1798 *Dictionary of the French Academy*, we see the term "esquisse" defined for the first time as the first trace a painter sets on a white canvas. However, when architects hear the term, they immediately think about the École or the École Nationale des Beaux Arts in Paris. In fact, it all began in the Middle Ages when art and architecture were not formally taught. Master architects simply passed on their design skills to young apprentices. In 1648, a group of young architects and artists, unhappy with their brotherhood of artisans, decided to dissolve that bond to create a new community—the first French fine arts institution in Paris. The group chose to follow Italian establishments of the time and, thus, began its classes as the Academie Royale de Peinture et de Sculpture.

Some years later, in the revolutionary spirit of the times, the academy's main goals were modified to provide free instruction and promote equality, making it possible for students from all socioeconomic classes to attend. The first embodiment of the École Nationale des Beaux Arts was born. Once the school was established, the task of defining a curriculum and corresponding pedagogy became paramount. Its founders decided that simplicity, grandeur, cleanness, and harmony were to be the pillars of the new edifice. Having no other points of reference, students were instructed to study works of art from Greek and Roman antiquity as a source of inspiration for their creative endeavors.

The Academie was composed of two sections: the Academy of Painting and Sculpture and the Academy of Architecture. In the painting and sculpture studios, esquisses in anatomy, geometry, perspective, and nude studies were assigned. In the Academy of Architecture, students began rigorous studies of Greek and Roman architecture, to be followed in the ateliers studios by architectural compositions that mirrored classical programs. This early tradition of applying methodologies employed in classical painting, and especially the esquisse, to the architecture design processes became a trademark of the school. It was a tradition that was to last for centuries.

Student discontent, climaxing in May of 1968, brought an end to the École's architecture arm. The Prix de Rome, another centuries-old tradition, was also discontinued. Today, only visual arts are taught at the historic building on 14 Rue Bonaparte.

PEDAGOGY AND THE ESQUISSE

Architectural instruction at the École was very much connected with the design processes of drawing, painting, sculpture, and engraving. Central to the propaedeutic part of the design process was the esquisse. Conceived in the same spirit as in the arts mentioned earlier, the esquisse became the design mechanism par excellence. In other words, the core of the pedagogy was the working methodology. With time, the esquisse evolved and reached true polymorphic qualities. First and foremost, it became an indispensable part of the design process, from the conceptual phase to the execution level. Additional functions included its use as:

- An instrument for the study of art and architecture
- An instrument to develop students' powers of observation
- An analytical and comprehension tool
- A mechanism for rigorous reflection
- A system to evaluate and improve work and research capacities

- A communication device for abstract ideation

- An architectonic hermeneutics instrument

- One of the purest forms of architectural representation

The most important function of the esquisse was its capacity to act as liaison between the theoretical and practical worlds. What could be esquissed theoretically could be modeled in an esquisse. Moreover, the esquisse offered architectural instruction to proceed pedagogically in both vertical and horizontal directions. In the latter capacity, it proved to be a superior instrument because of its complementary and indissoluble links with the architectural studio and architectural modus operandi.

In the ateliers, the esquisse was used in a variety of exercises:

- For students to discover their own resources, abilities, and weaknesses

- As a discovery mechanism in the architectural conception process

- As a way to master architectural representation

- As a way to study functional, plastic, and construction approaches

- As a study tool to understand architectural theory and historic architectural movements

- As a way to awaken curiosity and spark apertures in the students' ways of thinking

- As a way to synthesize urban qualities or lack thereof

- As a way to represent essential space and qualities

- As a way to represent space, its components, its reciprocal relationships, and its modes of association

In most cases the exercises included three phases: analysis, esquisse, and architectural project. (The themes ranged from minimal complexity to developing particular pedagogic objectives.) Usually, the first two years of architectural education were intended to familiarize students with architectural notions and notation using the esquisse. It was common to accompany the final project to be presented to juries with a variety of esquisses relative to the project. Some ateliers followed the juries with notes and commentary with formal pedagogical corrections. The critique of the esquisse and the choice of a solution constituted the most critical phase of the design process before the final stages got underway.

In the case of more complex architectural interventions, the process included four phases: analysis and programming; esquisse; an avant-project or preliminary solution; and a detailed project. In order to obtain their diplomas, students had to complete a series of short projects, a long project, urban projects, and a series of esquisses. There was no uniformity in the number of esquisses students had to complete, but it was obligatory to present an esquisse complementing all short projects. The upper classes had to complete a minimum of two 12-hour esquisses. In the latter years of the École, graduating students produced a variety of esquisses delineating their final project's design approach.

By associating the esquisse with the design process as a conception, representation, and communication mechanism, the École developed a pedagogy based in the arts and activities that sustain architecture as well as those thought processes that are educed from it. In this century, the process allowed design interventions at the very early stages of representation, before physical models, before photography, and before discourse. This allowed an awakening and the development of visual perception as a means to apprehend the real and to stimulate creativity and imaginary potential.

The closing of the École Nationale des Beaux Arts in 1968 did not stop the esquisse. Actually, most of the architecture schools that emerged at the time, as well as those that were in session and kept expanding, embraced many of the pedagogues of the École and especially the esquisse methodologies. A review of current curricula of these schools reveals a strong integration of the esquisse as the mechanism de choix in the design studio.

THE ADVENT OF THE COMPUTER

The original esquisse methodology used simple tools; paper and pencils sufficed. The results were satisfactory. When computers began to appear on the design studio horizon, there were many who thought that computers would eventually replace traditional pedagogies. At the time, many also believed that smart algorithms and artificial intelligence would free the design process from aesthetic constraints. After all, the computer was capable of generating and formulating an endless series of propositions under the guise of esquisses, which were the subject of analysis and choice. This led some to believe that the computer would be not only an ideal tool to present projects, but also a consummate tool to justify design decisions.

Following almost three decades of intense research, computer-assisted design (CAD) systems are still short of proposing a true tool, tool set, or tool kit to explore conceptual hypotheses while seeking spatial synthesis. Despite dramatic decreases in the cost of digital technologies making them available to many, systems have been used almost exclusively for the production of drawings and to communicate projects.

The central problem with current systems is that they demand a rigorous definition of all formal components at a time when projects are still fluid. Moreover, from a methodological point of view, following that process is contrary to the creative process. In the design studio, it dampens all creativity because students are asked to enter precise final geometries into systems at a time when spatial programs are still in the early stages. Architectural conception precedes production and communication—something not likely to change in architectural practice or in schools.

EMBRYONIC TOOLS

Because the computer design process is immutable and somewhat inflexible, if one wishes to integrate new tools into the procedure, one has to search for those applications that accept a certain conceptual fluidity in the same way the esquisse permitted designers to put their first creative acts on paper. Similarly, a new way to look at architectural representation is required. For these two prerequisites to have a real impact in the design studio, it is necessary for them to possess the attributes that follow.

1. Any computer-aided design technology or instructional procedure for esquisse or conceptual exploration must support the student in externalizing mental images in the most expeditious way possible. During this esquisse phase, the digital activity must assist the student in the following typological operations:

    • Plastic—gestural or sculptural form manipulation

    • Metaphorical—analog or symbolic associations with pertinent objects or circumstances

    • Useful—search for pragmatism and usefulness

    • Formal—phenomenological references manipulation

    • Functional—search for operational qualities

    • Materiality—exploration of haptic or tectonic qualities

    • Lighting—manipulation of architectural uses and effects of natural and artificial lighting

- Color—study of visual effects

2. The procedure must permit visualizing the evolution of the esquisse from the early conceptual phases to more refined representations.

3. Notwithstanding the hardware platform or software configuration, the direct esquisse on a computer monitor is only possible when the teaching methodology allows for sufficient flexibility and approximation in geometric construction.

4. Instead of spending considerable energy in hardware and software instruction, it is more effective to use a limited set of commands to esquisse and visualize architectural spaces.

5. During the esquisse phase, students must be able to easily access the three dimensional construct in order to modify it at every step in the design process.

6. From a single construct or model, students must be able to easily visualize it using a variety of representational techniques.

THE ELECTRONIC ESQUISSE

Following the previous considerations, the author has conducted several variations of the electronic esquisse (known to most students as the EE) studio in an attempt to conjoin the procedural and pedagogical directions established by the esquisse several centuries ago with current digital technologies. The EE studio has enjoyed ever-increasing success based on final outcomes of the studio and on course evaluations conducted at the end of the sessions.

Other factors come into play, too, such as the all-too-common dilatory attitude of some students. Procrastination is a word one hears often in design studios. It is not uncommon for students to wait weeks for a magical moment of inspiration. Also, when a project involves any kind of digital intervention, studio critics and computer instructors are often guilty of spending considerable time teaching highly detailed software commands. Complicating matters further, when a project is proposed within a specific time frame, there is a negative tendency to adjust the amount and quality of work by the perception of what can be accomplished in the last days or even hours before a project is due.

To resolve the issues raised above, the EE studio follows the École Nationale des Beaux Arts pedagogy as outlined at the beginning of this chapter, especially in regards to the way projects are organized. For example, while traditional studios explore a few projects—sometimes only one per academic session—the EE studio assigns, in the spirit of the École, 10 short projects within an academic quarter. This equates to a different project every week.

Timing is at the core of the approach. Instead of allowing students to wait for a magic moment of inspiration that in most instances proves elusive or to spend considerable amounts of time and energy learning obscure computer commands, the esquisse methodology stresses that the creative process needs to materialize as soon as the design challenge is posed. Once it starts, the computer becomes a wonderful aide because of its limitless transformational capabilities.

While the projects resulting from this methodology are obviously not as thorough as those produced under traditional schedules, the technique has enriched immensely the architectural experience of students in many ways:

- Instead of wandering for days and even weeks searching for an approach to a project, students are forced to concentrate on the major issues that the project poses. This applies to both design and computer issues. Due to the time limitations, priorities are set faster and in a more direct way. No time is lost on irrelevant design issues or complicated software instruction.

- Architectural concepts are communicated succinctly. There isn't time to waste exploring CAD schemes that have dubious results.

- As students are hard pressed for time, they quickly realize that other students may lend advice or know how to do things faster. Therefore, students interact better with one another as they learn that solo work can leave them without a support mechanism, which may be very much needed.

- The students' levels of confidence increase considerably because they test themselves in high-pressure situations.

EXPERIMENTAL PROJECTS

A predominant quality of students' studio work is its direct, experimental nature. Even though after the projects are assigned, some students choose to use the traditional, non-electronic esquisse as a point of departure, the first task is directed at approximating mental representations with the computer. EE studio's project typology is organized according to the following central themes:

- Form generation in two and three dimensions using combinatorial operations—testing spatial hypotheses

- Transformations as form generators

- Virtual reality modeling—variations on the same theme

- Solid modeling—the use of Boolean operations with primitives

- Exploration of spatial depth—visualization and rendering

PROCESS AND METHODOLOGY: AN EXAMPLE

Every project is introduced by a series of principles reckoning the journey from the mental image to the esquisse on the screen. At the beginning, only a few computer parameters and instruction sets are presented. With every new project, a new iconic scale is introduced to augment the conceptual representations. Also, a few more software commands are suggested in close harmony with the task at hand. The process continues by arriving at a construction method of a transmutable model. The idea here is to enrich every step by using the transformation capabilities of digital media. What follows is a typical one-week exercise that applies these concepts.

Students are first introduced to the fascinating possibilities that CAD provides in regards to performing a multitude of permutations on previously generated esquisses. The group is also made aware that second generation esquisses can become a great source of inspiration for new forms. In essence, the exercise consists of generating one basic three-dimensional esquisse and then conducting a series of transformations.

Parallel to the introductory remarks, students are provided with a set of basic generation and transformation computer commands. The given set of commands is restricted exclusively to those actions. Students are actually discouraged form using extraneous commands.

Students begin a quick esquisse, which can perhaps result in a simple series of cubes or trapezoidal shapes. The item can also be a very simple spatial representation of a notable structure, such as Mies's Barcelona Pavilion or even a Greek temple. At this point, instruction is provided on how to "walk around" this construct to see how it looks from various viewpoints. Students are constantly reminded to keep details at a minimum and to concentrate instead on the transformation possibilities. *This observation is critical because neglecting the former action usually leads to spending considerable energies on irrelevant software issues, while the latter tends to lead to new and exciting design schemes.*

Once the basic *esquisse* is constructed, students begin a process whereby the item changes gradually into another form. Transformations can be applied uniformly to the geometry of all the nodes of the item or can be placed only on selected objects, lines, polygons, or nodes. Students are then provided with instructions on how to use symmetry, translation, rotation and scaling operands.

The process continues by suggesting that topological replacement operations can affect elements, nodes, lines, and/or polygons where they: a) may have the same size and direction; b) may have the same size but a different direction; c) may have a different size but the same direction; and d) may have a different size and a different direction. The final phase of the exercise implies the use of one or more manipulative actions such as align, blend, convert, duplicate, replace, break, invert, divide, etc. At this point, students receive a list of possible actions and their corresponding software commands.

The final project, or rendu in the École, is submitted electronically via File Transfer Protocol (FTP) instead of by charrette, as in the days of the École. It consists of a least ten different transformations of the basic image.

HARDWARE AND SOFTWARE CONSIDERATIONS

Hardware manuals have become multi-volume texts with thousands of topics that are of little interest to the designer. CAD software packages have evolved from very rudimentary two-dimensional paint programs to systems with thousands of commands. These programs and systems now include hundreds of options, which are practically useless for architectural design.

In connection with this are two very common pedagogical mistakes in the subject disengaging the design process from computer instruction and placing an excessive emphasis on learning as many commands as possible. Both actions not only wipe out creativity but also make students lose a sense of what is really important in the design process.

Accordingly, the EE studio adopted the following premises on the use of digital media:

1. Instead of lecturing on the functions of various software commands, the pedagogy adopts a confrontational position. The idea is to enrich the digital experience by following the "less is more" approach. Students are encouraged to present and discuss different ways to reach the same point on a competitive basis—an experience that has proven to be not only very entertaining but highly effective.

2. In lieu of limiting the educational experience to software command teaching, the students are presented with a wide range of digital assists that allow them to:

   • Open and share their esquisses with fellow classmates not only to accept critiques but also to be able to determine common points and procedures

   • Explore, to perceive and to know the genesis of images and objects as well as the characteristics of the creative procedure

   • Explore endoptic, mnemonic, perceptual, conceptual, physio-perceptive, psychedelic, and imaginary objects

   • Explore the tactile relief, stereoscopic, stereographic, tri-dimensional, synesthetical, and physiologic attributes of images, objects, and spaces

   • Investigate iconographic, iconic, and verbal systems as well as semiotics, morphology, and the syntax and semantics of images, objects, and spaces

   • Address light and color as well as the thematic needs applied to pictorial or spatial research

   • Speak to the question of the civilization of the image and the architectural object

Acknowledgment

The illustrations in this paper are from students who have taken the Electronic Design Studio.

References

École Nationale Supérieure des Beaux Arts, The Renaissance in France: Drawings from the École des Beaux Arts, Paris, 1995.

École Nationale Supérieure des Beaux Arts, Paris-Rome-Athènes: Le voyage en Grèce des Architectes Francais aux XIX et XXe siècles, 12 Mai-18 Juillet 1982; Pinacothèque Nationale d'Athènes, Musée Alexandre Soutzos, 15 Octobre-2 Janvier 1983; The Museum of Fine Arts,

Houston, 1er Juillet-4 Septembre 1983; IBM Gallery of Science and Art, New York, 2 Février-24 Mars 1984.

École Nationale Supérieure des Beaux Arts, The Beaux Arts and Nineteenth Century French Architecture, Paris, 1982.

Nationale Supérieure des Beaux Arts, Pompéi: Travaux et envois des architectes français au XIXe siècle, Chapelle des Petits-Augustins, 12 Janvier-22 Mars 1981; Institut Français de Naples, I I Avril-13 Juin 1981.

Egbert, Donald Drew, *The Beaux Arts Tradition in French Architecture*, Princeton, NJ: Princeton University Press, 1980.

École Nationale Supérieure des Beaux Arts, Les Concours Chenavard (Section d'Architecture) à l'École Nationale des Beaux Arts, Paris, 1909.

École Nationale Supérieure des Beaux Arts, Les grands prix de Rome d'architecture de 1850 à 1900; reproduction en phototypie del lers, 2mes, et 3mes seconds grands prix avec les programmes des concours, sujets donnés par l'Academie de Beaux Arts. Paris, 1900.

CHAPTER 26

Teaching Structures Through Multimedia: An Interactive Exploration of Structural Concepts Using Digital Technology

Shahin Vassigh

Assistant Professor

State University of New York—Buffalo

Buffalo, NY

INTRODUCTION

Developing innovative approaches to education is not unique to architecture. However, the technical component of a creative degree program, developing innovative approaches to teach structures within the architecture curriculum, is not only desirable but also absolutely necessary. Currently a large number of architecture schools across the country are engaged in teaching structures using a traditional engineering pedagogy. The fundamental problem is that although understanding structures lies at the core of the education of the architect, the traditional engineering-based approach does not facilitate an intuitive and conceptual grasp of the material. Therefore, the subject is considered to be too difficult, uninteresting, or simply intimidating for many students. Uninspired by an overly technical approach to structures, many architecture students look at structures as a necessary requirement for graduation and a subject that is often stale and sometimes incomprehensible. Many question whether or not they will ever use what they have been taught.

Architects think, learn, and approach the design of the built environment differently than engineers; however, understanding structures and structural behavior of built systems is critical to the design process. If architecture students are to effectively learn and then apply sophisticated structural analysis and design, the teaching methods used must respond to the needs, capabilities, and perspectives of the architecture student. Advances in computer software, improvements in graphics capabilities, and the shrinking cost of producing high-end graphics, data, and hypertext allow the development of new teaching tools that more effectively communicate both basic and sophisticated structural analysis and design concepts to architecture students.

ABSTRACTION, REDUCTION, AND THE SCIENTIFIC METHOD

Given the fast rate of progress in structural science, combined with the increasing need for specialization within building design and construction, engineering has gained enormous authority over the material sciences, construction technology, and the design of the built environment. The growing role and influence of the engineer has lead to the introduction of sophisticated mathematical models into the building construction process. As a result, the scientific method and mathematical rationalism have become the dominant models for teaching structures to both engineering and architecture students.

The problems with the current approach to teaching structures in most architecture programs are rooted in the nature of scientific inquiry. The scientific method, by its nature, is abstract and reductive, and it seeks to develop a quantitative explanation for all events in the physical world. Although abstraction and reduction are to find analytical solutions to structural problems, understanding structural behavior through component analysis and behavioral formulas creates serious impediments to developing a comprehensive understanding of structures. The analytical method of understanding structures mandates consecutive dismantling of the structure into to its components, detaching it from all the connected members, then focusing on a particular element that is further reduced into structural symbols and annotations. At this point, the structural member would be prepared for the application of static equilibrium formulas, which will lead to a quantitative determination of the member forces, which will be used in conjunction with other member information in future analysis.

This process disconnects all the conceptual and visual relations between a structural member and the actual building structure. In most cases, a complete picture of the structure is never presented prior or after the analysis. The engineering-based model also seldom engages the overall structural system to clarify the relationship between the small-scale components and the large-scale structure. This model facilitates operating on a constructed hierarchy that is based on an orderly study of beams, columns, trusses, and connections—all studied in a separate context from the structural system.

AUGMENTING THE SCIENTIFIC METHOD WITH DIGITAL TECHNOLOGY

This critique of the scientific model does not deny its immense value in teaching structures and providing a practical framework for analyzing structural behavior. No qualitative method can replace the knowledge gained by the scientific method of analyzing structural behavior. However, given the very different needs and capabilities of architecture students, how can one approach teaching structures within the scientific framework without sacrificing conceptual and intuitive aspects of understanding structures, which in many cases lead to more interesting and creative design?

The evolution of computing technology over the past decade provides the means to develop very different, innovative solutions to the problems of teaching structures to architecture students effectively. The conventional use of computer-based structural analysis utilizes the quantitative speed of computers; however, advances in graphics processing have enabled relatively inexpensive, readily accessible computers to develop, produce, and display powerful simulations of reality and create experiences that are physically intangible. Digital technology enables us to see what we are not normally able to see. The new technology can fabricate and reconstruct conceptual and practical elements lost in the process of reduction and abstraction.

I have been developing an instructional software package that utilizes a wide range of graphics, animation, sound, video, and other interactive devices to communicate the principles and application of structural analysis. Written using Macromedia Director, an HTML composition and presentation development tool, the completed package will be available on CD-ROM or directly accessible over the Internet.

At its highest level of organization, the program divides the study of structures into three areas.

1. *Basic Concepts:* general structural analysis and design concepts, definitions and working principles

2. *Structure Systems:* a searchable database of structural subsystems (trusses, cables, arches, beams, etc.) and how they behave and are modeled mathematically

3. *Building Structures:* detailed examinations of large-system structural behavior, studying the work of great architects through interactive images that examine structural behavior as a whole or detail a specific structural sub-element

The program radically differs from traditional textbooks and media in three ways. First, the graphic images used are three-dimensional. Most are animated to directly demonstrate structural behavior under conditions of structural stress. Rather than being abstract representations of structural components, these images are highly detailed real-life images of beams, columns, trusses, etc.

Second, through the use of linked menus and hypertext, all three areas of study are linked and accessible from within each other. This process offers the possibility of studying each structural element in a separate layer or fusing many layers of information into a single image. At each level of this process, a general overview of the structural system provides the visual axis grounding the investigation, but successive layers of information (mathematical formulas, analytical results, graphic representations of behavior, etc.) can be accessed and overlaid onto the buildings, issues, and sub-systems currently being studied. This process provides a holistic approach to the investigation of structural concepts.

Third, the program is considerably different in look and feel than traditional texts and study supplements, particularly those of engineering. The overall approach is to provide a presentation and analytical format that is more familiar to the thinking and graphic communication conventions of the architect, as well as provide direct insight into the design of known works of architecture. The system will include voice, text, and animated interviews with the great architects whose works are included for study.

A brief example demonstrates how the program works. Figure 26.1 shows the beginning menu for starting the program. Selecting "Building Structures" from the menu, create a link to the next menu (Figure 26.2), which brings up a list of the architects in the database. By choosing an architect from the list, such as Renzo Piano, the link to the next file will be activated. This file (Figure 26.3 on page 340) presents a list of buildings designed by Renzo Piano available in the database. After selecting a building from this list, for example, Unesco Laboratory Workshop, an overall image of the building will lead to an interior view, showing all the exposed structural members (Figure 26.4 on page 340).

Clicking on a beam, a column, or connections triggers the proper link. For example, if we select a beam from this image and press on the text icon, a general description of the beam member will appear (Figure 26.5 on page 341). Clicking on the

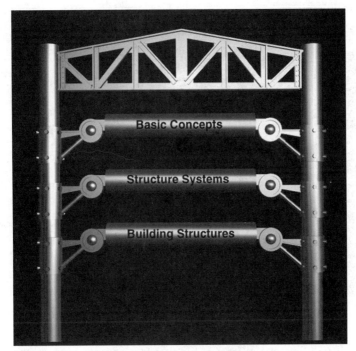

Figure 26.1 *The beginning menu*

Figure 26.2 *A list of architects in the database*

Figure 26.3 *Buildings designed by Renzo Piano*

beam again will link to an animation file, which graphically explains the load-bearing mechanism of a simply supported beam. The following files will explain the beam action in response to vertical shear (Figure 26.6 on page 341), horizontal shear

Figure 26.4 *An internal view of the Unesco Laboratory Workshop*

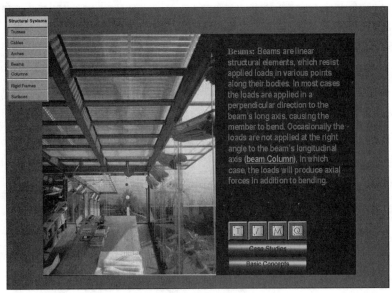

Figure 26.5 *A general description of the beam member*

(Figure 26.7 on page 342), and a combination of the two, which is the beam's flexure (Figure 26.8 on page 342). At any point, the pop-up windows provide choices for connecting to the relevant structural concept or structural system menus.

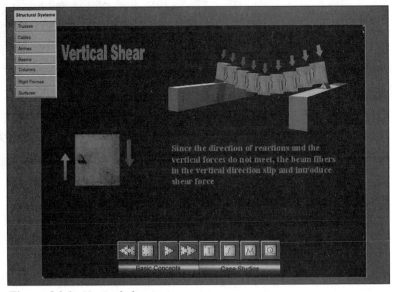

Figure 26.6 *Vertical shear*

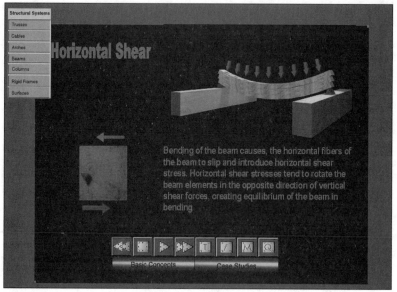

Figure 26.7 *Horizontal shear*

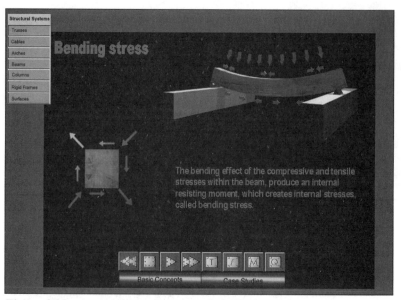

Figure 26.8 *A combination of vertical and horizontal stress*

ADVANTAGES AND OPPORTUNITIES USING DIGITAL STRUCTURES

The use of this technology offers a number of advantages to teaching architecture students the theory, principles, and application of high-level structural analysis and design.

Visual Enhancement

One of the best ways to convey concepts and information to students is to channel instructional concepts through familiar media and information formats. Architecture students are visually oriented and have extensive training in three-dimensional media. Most often, two-dimensional textbook diagrams do not engage their full attention. Taking advantage of the architect's visual language to describe the structural behavior can be very effective in conceptualizing structural behavior.

Contextual Treatment of the Subject

A significant portion of architectural education is based on case studies and examining the work of various architects and landmark buildings. Teaching new concepts in a familiar context can facilitate the introduction of new subjects and bring their understanding of the building into a new light by explaining how the building works through structural analysis. Using digital technology can provide the option of looking at a specific structural system instantaneously linked to the work of various architects or to specific buildings. For example, the subject of rigid frames can be studied in the work of architects such as Mies Van de Rhoe or Corbousier, or a building such as Falling Water. Starting with a general image, the building is introduced, then through de-layering or skinning off the building through image processing, the structural system under investigation will appear. Presenting careful choices will give students the option of studying structures within an architectural context, a medium that they are very comfortable with and know well.

Capturing Interest

One of the great challenges in teaching technical subjects is creating a high level of interest and engaging the students in the subject fully. In the case of architectural education, the curriculum revolves around the studios as core courses, which are clearly far more interesting and exciting for the students than the dry technical courses. Interactivity, animation, sound effects, and video clips are among what digital technology can bring into the picture, adding entertainment to what is normally considered to be boring and stale.

Multilevel Readings

This approach provides an opportunity to study the same subject in a variety of contexts. The students can access the subject under investigation in the exact form delivered in class, receive instruction, repeat information, and use multiple views and magnifications to investigate the subject more thoroughly. Selecting a different path on the menu, it is also possible to avoid the repetition and study the same subject in a completely different context. The structured self-exploratory system allows the students to explore concepts, from introductory to advanced, at their own pace.

Facilitating Advanced and Alternative Design Concepts

The graphic manipulation of the real-life pictures can serve as an important teaching device. The example here presents a screen shot from the lesson in rigid frame, the corresponding deflection mechanism, and the moment diagram under the application of live loads (Figure 26.9).

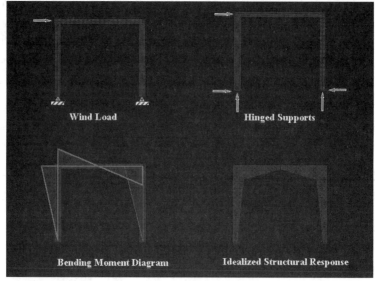

Figure 26.9 *A lesson in rigid frame*

Typically, studying the variation of internal forces and moments along structural members and the mechanics of structural deformation are among the most important components in teaching structures. However, in most cases this is done through quantitative analysis, with no application in architectural design. Figure 26.10 shows a building designed by the architect Santiago Calatrava. To give a practical sense to the utility of the moment diagram, the moment diagram of the bending frame is drawn directly on the building frame.

Figure 26.10 *A Santiago Calatrava building design*

The moment diagram superimposed on the photograph of the frame explains the structural logic of the frame form. As seen in the photograph, the frame has its maximum mass at the corners where the moment is maximum. The mass of the structure becomes minimized at the center of the frame, responding to minimal moment and stresses at that location. The manipulated photograph in itself provides a strong example of the utility of the moment diagrams, which are standard procedure in structural analysis. This can change the students' conception of drawing moment diagrams from being a useless effort to acting as a possible architectural design tool.

Availability and Distribution

The first step in the introduction of this software is to make it available on CD-ROM and directly over the Internet. The possibilities for expansion of the software

are truly unlimited. Periodic modifications can add modules about new building structures, new technology, and new materials. Another set of modules can incorporate exam aids, tabulated information, and interactive example problems.

CONCLUSION

This chapter suggests an unconventional method in teaching structures that can serve as the first step in bridging the gap between teaching and learning structural concepts effectively. Through visual and conceptual enhancement of the introduced subject, students can improve their understanding of the course material. This approach can also bring new levels of interest and excitement to the subject matter, changing the way structures is treated within the architecture curriculum, which in turn can facilitate the integration of structures within the rest of the architecture program. Another positive impact can be the integration of the students' technical understanding of the structures into their studio designs. Learning structures as an integral part of the architectural design, rather than as a separate and additive step in the design process, can provide the basis for designing structurally efficient, serviceable, yet expressive buildings and a method for professional application once the students leave the academy.

Photograph Credits

Figures number 4 and 5: "Renzo Piano building workshop," published by Phaidon Press Limited, London, England: 1995.

Figure number 10: "Photo EL CROQUS 38+57 Santiago Calatrava 1983-1993," published by EL CROQUS, Madrid, Spain 1985.

CHAPTER 27

Multimodality Concept Maps and Video Documentary Reconstructions: New Uses for Adaptive Multimedia in Learning

Roger D. Ray
Professor of Psychology

Rollins College
Winter Park, FL

Growing and compelling constructionist learning literature has made "active learning" a touchstone technique for many of today's educators. Likewise, concept mapping has become a widely respected assessment strategy for evaluating how students articulate the meanings they construct from such personal experiences. Unfortunately, the extant literature on concept mapping remains focused on the technique's origins as a language-based tool. For example, Novak and Gowin (1994) describe only the use of words, phrases, and qualifiers in pursuit of understanding a student's linguistic associations within a topical domain. Nevertheless, it is possible—even desirable—to articulate a broader view of concept mapping as a multi-modality knowledge representation activity.

THERE IS MORE TO CONCEPTS THAN WORDS

Closer study of knowledge representation and communication processes reveals sorely neglected possibilities for expanding the concept mapping tool set beyond the singular use of *linguistic representation*. One outstanding example is evident in Tufte's (1983, 1990) seminal illustrations of the power of *graphic representation* for conveying complex concepts and events, such as temporal and geographical landmarks attending troop losses during Napoleon's devastating Russian campaign. Mathematicians, engineers, and even musicians would quickly add *symbolic* (e.g., mathematical) *representation* schemes to the arsenal of conceptual representation and communication. And I cannot imagine an architect trying to convey the nuances of a new skyscraper concept to a civilian client without the use of physically or computer-constructed *simulations* or *models*. Unlike two-dimensional blueprints or elevational drawings, the construction of a simulated model allows direct contact with a building's three-dimensional space and proportionality, not to mention allowing a direct representation of the way the building is likely to impact the visual landscape surrounding it. Imagine the limitations if such a conversation between architect and client were constrained to words alone!

Anyone dealing with complex events and objects will quickly appreciate the ability to communicate their thoughts and knowledge by using a variety of expression modalities. At a minimum, words, graphics (including line drawings, photographs, paintings, movies, and even geographically arranged information such as tables), symbols, and constructed models will find their own power for communication and illustration.

KNOWLEDGE DIMENSIONS BEYOND REPRESENTING AND ASSOCIATING

Just as there are other modes of expression beyond language, there are also other dimensions to knowledge that have been ignored by those who stress concept mapping-dimensions that play a critical role in defining the broader scope of knowledge and reasoning skills. In the section above, I introduced the idea of alternative *expression modalities* and stressed the act of event and object representation through a given mode of expression. But the concept mapping literature stresses, albeit within the linguistic mode, the act of association. That is, during concept mapping exercises, students are asked to associate words with other words. For example, when one hears the words "John F. Kennedy," it is almost impossible for those of a given generation not to think also of "United States president" and "assassination." These are ingrained associations between words or phrases.

But we seldom recognize a need to evaluate associations in other domains. Nevertheless, within-modality association is also at the core of solving mathematical equations, whereby one equivalent form of an equation is generated from another, as when $d = rt$ is associated with the alternative, but just as accurate, forms of $d/t = r$ or $t = d/r$.

Likewise, how many problems can you think of which require you to associate between graphic representations? If given a painting by Monet and a multiple choice of four other paintings by various painters, only one of which is another Monet, could you identify the correct one? Any art historian or museum curator almost certainly could. But how often do we strive to teach our students to rely upon such non-linguistic association skills? And how often do we use non-linguistic associations for evaluational concept mapping? If inspection of the research literature is any indication, not very often. But this only emphasizes just how much in today's educational processes we stress language to the exclusion of other modes of knowing, and thus other modes for expressing our knowledge.

However, there are even more dimensions to knowledge than representation and association. For example, one can ask students to *translate* one modality of expression into another. Examples might include asking a student to verbally describe the meaning of a graphic data representation, or asking a student to draw a graph, which illustrates the possible values of a (symbolic) linear equation. Mathematics teachers do this everyday, of course. But how often do we teach translation skills in other disciplines? It is a skill that is seldom articulated as an educational goal in and of itself, despite being observed within some disciplines on a regular basis. I contend that teaching such skills on a broader scale is ignored because the teaching profession lacks a clear taxonomy of the complete skill set students require for knowledgeable communication.

Translation between modalities is a skill, which can be characterized as making associations *between* alternative modes of expression, as opposed to making associations only *within* a given mode. For example, musicians not only write music, but also perform it from the written score (a type of modeling). Again, mathematics is exemplary when students are taught to translate problems presented through words into appropriate formulas. These formulas then lead, through processes of *within-domain associations*, to concrete solutions to the verbal problem. It is also relevant to reflect at this juncture on how often math students are encouraged to "draw a picture" of a word problem as an intermediate translation to help generate the correct mathematical formula.

I have personally noted, however, that skills taught in mathematics courses do not always generalize immediately to other disciplines. For example, it is surprising how readily a student will verbally interpret a "rate of response" graph correctly in psychology but still be unable to, say, tap a table at that same rate. This is a problem involving graphic-to-modeling modality translation, and it is a very poorly developed skill in most students who aren't engineering majors.

But perhaps the most powerful knowledge dimension comes when one both translates between and associates within modalities simultaneously, thereby constructing what I refer to as *aggregated concept clusters*. An example of such aggregation was just accomplished—I hope!—in the above paragraphs when words and mathematical formulations were used in concert to generate the concept of *across modality translation* as a form of knowledge communication. And that discussion presumably has much more power for the audience who also will see it modeled through a software system that embodies its principles than for those who are restricted to only reading it. The demonstration models the aggregation of various domains, not just the words. As such, the software represents my own aggregated concept structure and it includes all domains-verbal, symbolic, graphic, and constructed models-in concert at the same time.

But how many of us in the teaching profession set out to include the specific use of all such concept expression tools when we aspire to teach knowledge construction skills? And why not? I have already suggested that one limitation has been the lack of a conceptual taxonomy of component skills. But there are other limitations as well, including the lack of physical expression tools and the problem of defining appropriate problem spaces in which to develop a student's skills.

A Multimedia Problem: The Complexity of Knowledge Expression Tools

While some teachers simply miss seeing the value in teaching multiple modality thinking and expression skills, others are convinced of its relevance. Nevertheless, it still is not an easy task to assume. Besides lacking a taxonomy for guiding our goals, a second impediment has been our inability to integrate physically such an array of expression modalities. For certain, some teachers involve students in class projects where many modalities of expression and construction can come together in project constructions, as happens in engineering fairs for example. But rarely do we emphasize multiple modality expression as a matter of everyday study skill development. The difficulty of managing such a multi-modal learning process is not trivial. Tools for thinking and communicating through integrated multiple modalities of expression have not been very user-friendly until very recent times. But as multimedia technologies advance, those advances offer new opportunities and challenges for teaching in significantly new ways.

I have taught a sufficient number of years to actually remember when we faced a technological limitation in teaching students to practice alternative word and phrase constructions when we assigned essays or term papers. Some in my audience may remember with me when teachers specified the use of a typewriter to ease the burden of reading these productions once they were submitted. But most teachers hardly expected a paper to go through rewrites based on editorial feedback on multiple drafts. Retyping after physically cutting out words, lines, and paragraphs and subsequently pasting them back together for editing purposes was just too laborious to be practical.

Then, computer-based word processing arrived. As noted above, cut and paste is more than an electronic metaphor-it is how we used to actually do it. And given today's word processing technology, many more teachers now expect multiple drafts of essays and term papers. These teachers now rely on rewriting as a means for teaching critical editorial skills-skills requiring one to reflect upon writing structure and choice of wording. This brings the teaching of writing to a whole new level; and it should be stressed again that this revolution has developed as a result of our incorporation of computing technology into our teaching. Many pundits who lament the stress on technology in teaching seem to have missed the significance of this non-trivial application of computers in our everyday assignments.

Technology is being used even more significantly by some. Many of today's more adventuresome users of computer technologies in support of the teaching-learning process have emphasized both individual and team-oriented student productions of multimedia "portfolio productions." Multimedia technology has changed opportunities for expression in a very fundamental way. It is now possible to have students expressing themselves in one highly integrated presentation without having to construct or collect anything but an array of alternative representations in different modalities. After collecting relevant component assets, students may then connect them into complex arrays of hyperlinked media associations.

A growing number of technologically sophisticated teaching pioneers glimpse intuitively many of the reasons for such technology applications. Like those before them who saw the multiple editing and rewriting advantages inherent in electronic word processing over hand-written or even typed, single-version theses, these multimedia-oriented teachers see practical, if not theoretical, advantages of teaching students to express themselves through a multitude of modalities.

The use of multimedia authoring is constrained by its seeming complexity. Even teachers who get intuitively excited and motivated to use multimedia tools in their teaching are all too often discouraged by extant authoring technologies. These limitations, of course, lie with the software technologies, not the hardware technologies. Thus, many teachers, and to a lesser extent their students, are all too often intimidated by the inherent complexity of the authoring task. This is the dilemma I first faced when, despite my understanding of why I wanted to use student-oriented multimedia, I had to figure out how I could use it. That, and a number of other contributing factors, led me to develop both a new software system, with relatively simple authoring attributes, and a comprehensive strategy for incorporating its use into my teaching.

A Second Multimedia Problem: Complexity of Aggregated Concepts

I have noted that a significant problem impeding multimedia-teaching efforts is the lack of sufficiently simple authoring environments that won't detract unduly from the primary production and expression task. But another problem of almost equal importance exists as well. That is the problem of helping students find a sufficiently defined "problem space" to explore without their losing sight of the intellectual growth that comes from depth as well as breadth of exploration. Let me cite an analogy to express this problem.

Have you ever reflected on how abstract and disconnected your course activities are likely to be for beginning students in their first week of classes? Despite your best efforts to give a preview or outline of where you will be going and what you will be doing during the semester, the lack of meaning in technical terms, the students' inability to follow your sophisticated representations, associations, translations, and especially the aggregations that you will construct with or for them over the semester makes previews a somewhat futile exercise. If you don't believe me, follow up a preview session with some in-depth, concept-mapping interviews with some of your students. It isn't a reinforcing exercise for instructors.

So how do we get a student to better appreciate the overall picture of where we would like them to go in their next few months of intellectual growth and pursuit? And how does this relate to multimedia authoring? Despite the case for multimedia authoring by students, when students are first exposed to hyperlinked libraries of information, they are quickly and easily overwhelmed. In fact, one of the most commonly cited problems associated with the use of hypermedia libraries is that of getting lost in the web of information. Hypermedia libraries lack a coherent integrative mapping of their purpose and content, much less their inherent integration of content elements.

SEEKING SOLUTIONS TO MULTIMEDIA TEACHING PROBLEMS

Over the past 10 years or so, my laboratory has been addressing both problems described above (i.e., lack of accessible multimedia authoring tools and defining clearly stated and meaningful problem spaces for creative expression). We started first on authoring tools by building non-scripting templates for authoring multimedia presentations in Apple computer's Hypercard. That work evolved and we now have both cross-platform and Internet versions, which incorporate highly sophisticated tool sets. The software system that grew from these efforts is called MediaMatrix.

Beginning about this same time, I also started exploring the other multimedia problem by using documentary videos as a means for motivating some independent-study students to explore a relatively finite, yet sufficiently open, "problem space." I asked students to conduct what I call "production autopsies" of NOVA or *National Geographic* productions.

Video documentaries offer highly sophisticated resources for telling a complex story in a relatively short period of time. It is quite likely that the very origins of human cultural transference from one generation to another lie in the stories we are told. Whether handed down through traditions of oral narratives, pictorial archives-like those seen in ancient Egyptian tombs and even more primitive cave paintings-or written down as recorded linguistic historical accounts, our sense of culture comes largely from stories. We easily recall story lines. They make sense because they have a theme and a temporal dimensionality. It is almost as if we begin our lives being told stories and end them telling stories. So if you wish to quickly submerge a student into a very complex topic, as most courses intend by their titles, what better means for doing so quickly, vividly, and dramatically than by a documentary video that summarizes the content and importance of your course? That is a strong argument for using video previews.

In my "film autopsy" assignments, students demonstrate their individual or collective connections among ideas presented within a given documentary by reconstructing much of the substantial research that was required on the part of the authors and producers to bring their documentary to fulfillment. Students then archive their research products and hyperlink them to appropriate components of the video documentary. I refer to this as building a text-and-graphic "back-end database" to the digital video "front-end" presentation.

Let me actually demonstrate this concept with a very brief excerpt from a BBC (British Broadcasting Corporation) production called *Discoveries in Animal Behavior.* Originally shown in the United States as a part of the Nature series on PBS, this multi-hour series reviews many of the historically significant discoveries of behavioral principles through reconstructions of landmark experiments. Today's demonstration is a selected singular clip on E. L. Thorndike's discovery of his "Law of Effect" concerning the power of reinforcing consequences in the process of skill learning. We begin with the video itself, viewed as a digital, full-screen presentation. At the bottom of the screen is a running caption of the spoken narration, with selected words bolded to indicate their hyperlinkage to other associated elaborations or illustrations. Likewise, if at any time the video is paused, hot spots within the video image are highlighted and labeled to indicate that the video, and not just its narration, is also hyperlinked to multimodal expansions.

Importantly, the MediaMatrix software system allows one to connect a multitude of related objects, including full textual essays, brief pop-up footnotes, still pictures, relevant video comments, other conceptually related full-screen video topics, tabular summaries of relevant data, and even full-scale laboratory simulations that illustrate the concepts to which they are attached. This is most easily demonstrated by navigating to a textual topic, say a biographical essay on Thorndike, and shifting the software into authoring mode. Making any selected text bold, to indicate that it is a navigable hot link, also allows us to connect an entire matrix of associated topics, comments, videos, tables, or simulations.

THE RELEVANCE OF FOUNDATIONS TO AGGREGATED CONSTRUCTIONS: THE ROLE OF TUTORIALS

It is unfortunate, in my opinion, that those who stress contructionistic learning all too often ignore the relevance of foundations. As with physical buildings, the foundation of any construction determines the overall integrity and strength of the more obvious elements. As this metaphor relates to knowledge systems, foundations are typically defined by the representation repertory of the student-the technical vocabulary and even the jargon used within a discipline of study. All too often, if a student lacks a sufficiently developed sense of fundamentals, the discipline itself becomes meaningless as it expands in associative scope and aggregative structure. But how are foundations best laid?

The complete answer to this question is complicated and, therefore, beyond the scope of this chapter. Suffice it to assert that one very important component to establishing a sound knowledge foundation lies with a student's ability to map real events to relevant representative expressions-whether they be descriptions, graphic illustrations, symbolic formulations, or exemplary models-which describe those events. And very often, this representational process, and even some associational processes, must be repeated many times in order to establish sufficient generalization to make it truly meaningful. I strongly believe that computer-based tutorials can play a significant role in this early developmental process, especially if tutorials can automatically adapt to the individual needs of a given student, as human tutors are apt to do.

As such, the MediaMatrix software system was also designed to be an artificially intelligent, adaptive instructional tutoring system. If we switch modes of use and initiate the "Tutor" mode, we find ourselves in a tutorial system that adapts level of difficulty and even content of questions to meet the differing needs of each individual viewer. Based on an imbedded automated knowledge generation engine, MediaMatrix not only tracks the user's path of use, but also develops a mirror image of his/her developing knowledge network of associations as measured by the tutorial questions. This is made possible by the fact that all questions are coded for their incorporation of at least two concept terms, images, or actions. As a student develops an understanding of these associational relations, answers to questions reflect which associations are being constructed. This mirror image of the student is then used to select the type of presentation, form, and difficulty of the tutorial question to be asked, and even which content the question should stress. This is a process called *adaptive instruction*.

It is beyond my present purpose to detail much more about the adaptive instructional concept itself. But those interested in knowing more about how it works, why it is relevant to tutorial processes, and which pedagogical principles drove its development might want to read other articles on the topic (e.g., Ray, 1994; 1995a; 1995b; 1995c; Ray, Gogoberidze, and Begiashvili, 1995). Once I saw the fundamental merit of the multimedia approach, I immediately saw the need for building simpler, yet more powerful, authoring environments.

THE ROLE OF MULTIMEDIA AND TUTORIAL AUTHORING

I especially wanted to allow students to focus more on the content and connectedness of their production efforts, not the technology per se, which I feel is certainly important, but should remain secondary. That is, I want to use technology, not *teach* it. I sometimes, even somewhat sarcastically, refer to the teaching of computer technologies as being equivalent to the offering of a curriculum on the typewriter 50 years ago. Nevertheless, I do appreciate the rather significant differences between the computer and the typewriter. And this isn't to overly demean typing skills. I have always appreciated being allowed to take a course that developed my own typing skills while I was still in secondary school. I just would prefer that others teach computer technologies while I get on with teaching my own primary discipline-psychology.

But as noted earlier, I also saw a need to do a bit of content authoring of my own to develop multimedia-based tutorials. I especially desired to construct sophisticated adaptive instructional tutorial services for my more needy and struggling students, especially those in introductory psychology and statistics. Therefore, I set out to design an authoring environment that met both of my needs-easy to learn and use for authoring, but extremely powerful and incorporating artificially intelligent adaptive learning in its products.

I believe the ideal system should allow one to author either simple multimedia productions or complex tutorials, as the applicational ends dictate. I also anticipated that I would want to teach advanced students how to write tutorial services for slower students. This allows one group of students to reflectively construct learning opportunities for other students more in need of building conceptual foundations. Building tutorials also encourages student authors to reflect upon their own productions in highly constructive and critical ways, such as deciding which ideas are worthy of the development of questions. And if my own adaptive tutorials emphasized learning verbal and multimodality associations, then the student tutorial author would also be retrospectively "concept mapping" his/her own production.

The authoring tool we eventually created remains in constant refinement and evolution in my laboratory. As demonstrated, MediaMatrix allows students to include textual essays as well as original video documentary material as a fully integrated system. Superimposition of navigable hot spots within that text or video illustrates a student's conceptual linkages to illustrations, expansions, or otherwise associated materials. Given that students may branch to as many choices, incorporating as many varieties of media, from each "source object" as they desire, the system allows them to construct an aggregated matrix of their multi-modality associations.

Work in progress will eventually allow teachers quick access to a summary of these various linkage matrices, thereby offering a self-reporting summary of the students' mapping of connections among various types of content and media as well as specific topics. Tutorial questions based on cross-modality associations and even video element identifications are also possible. Of course, finding and developing the electronic assets for this expression, whether via traditional library and archive searches or the Internet, are integral components in the preparation of the final composition.

CONCLUSION

I have suggested that constructionistic learning can benefit greatly from modern advances in digital technologies. I have also suggested what I believe are sound pedagogical reasons for incorporating these technologies. The literature on concept mapping is highly restrictive in its exclusive emphasis on language as the means by which we express or evaluate knowledge. Nevertheless, concept mapping points the way to a much more elaborate and sophisticated approach to knowledge expression and evaluation if we add graphic, symbolic, and simulated modeling to our linguistic arsenal of production. Likewise, we need to understand the context in which concept mapping's emphasis on associational linkages is defined. If we view knowledge as incorporating representation, across-modality translation, and multimodality aggregation, as well as the traditional emphasis on within-modality association, then we begin to appreciate the inherent power of hypermedia constructions.

But constructing hypermedia resource libraries is a daunting task if one does not have a clear organization mnemonic to keep it confined and organized. I have suggested that one of the most motivating and powerful mnemonic tools one can use is that of documentary videos. Students who reconstruct some of the archival and constructional activities behind the director's and editor's voice find unique opportunities for building new associations, modality translations, and even aggregations that give the documentary depth and breadth it never had as a singular production. In doing so, students quickly learn meta-knowledge construction skills as direct goals of the enterprise. Technology, both hardware and software, is rapidly approaching the point where such pedagogical processes should be the rule, not the exception. But that presents quite another challenge-the challenge of changing teaching strategies and behaviors. Time will only tell whether or not we can eventually meet the challenges inherent in this application of our technological tools. We definitely don't have to wait for compelling pedagogical reasons or user-friendly tools any longer.

References

Novak, J. D. and Gowin, D. B. (1994). *Learning How to Learn*. New York: Cambridge University Press.

Ray, R.D. (1994). Using virtual reality to deliver laboratory experiences in undergraduate education. *Proceedings of the Orlando Multimedia '94 Conference*. Warrenton, Virginia: Society for Applied Learning Technology.

Ray, R. D., Gogoberidze, T. & Begiashvili, V. (1995). Adaptive computerized instruction. *Journal of Instruction Delivery Systems*. Summer, 28-31.

Ray, R. D. (1995a) Adaptive computerized instruction: Learning what students learn while learning, teaches computers to improve teaching while teaching. *Proceedings of the Orlando Multimedia '95 Conference*. Warrenton, Virginia: Society for Applied Learning Technology.

Ray, R. D. (1995b). MediaMatrix: An authoring system for adaptive hypermedia teaching-learning resource libraries. *Journal of Computing in Higher Education*. 7 (1) 44-68.

Ray, R. D. (1995c). A behavioral systems approach to adaptive computerized instructional design. *Behavior Research Methods, Instruments, and Computers*. 27 (2) 293-296.

Tufte E. R. (1983). *The Visual Display of Quantitative Information.* Chesire, Connecticut: Graphics Press.

Tufte, E. R. (1990). *Envisioning Information.* Chesire, Connecticut: Graphics Press.

CHAPTER 28

Using Technology to Enhance Student Learning in the Laboratory Through Collaborative Grouping

Anne J. Cox

Assistant Professor of Physics

Eckerd College

St. Petersburg, FL

William F. Junkin III

Professor of Physics

Erskine College

Due West, SC

INTRODUCTION

The authors have developed a strategy to improve student learning in the laboratory by pairing groups for brief discussions during the class. Specifically, the authors ask students questions via networked computers to probe their current understanding of material. The students' answers then serve as guides for the pairing of laboratory groups for further discussion. Focused student discussions lead to an increase in student learning, critical thinking, and communication in the laboratory. Using the networked computers allows the instructor to know when such student discussions might be more effective. We will demonstrate the strategy we developed and the software used in our introductory physics laboratories.

Colleges and universities increasingly are investing in information technology infrastructure, anticipating an increase in efficiency, i.e., less cost and/or more students learning more things. Distance learning and Web classes can reach more students, but some faculty members argue whether or not students actually learn more. In other words, how can or does technology enhance student learning? After all, converting a set of overheads to PowerPoint presentations, although more colorful, does not generally change a lecture dramatically. Using technology to collect more and better data in science laboratories can free students to spend more time thinking about the experiments instead of tediously collecting data. However, technology can also enable students to pay less attention to the experiment, think less about what was happening in lab, and leave lab earlier.

We sought a solution that was easy to implement and that effectively enhanced student learning in the laboratory. That solution involved a different use of technology in the laboratory. Specifically, we were interested in using technology in a manner consistent with current understandings of the importance of actively learning pedagogical strategies. What follows is a description of the technique we developed to enhance student learning in the laboratory using technology and some preliminary results. In this technique, the technology actually increases human interaction: Students talk with each other (face-to-face) more when we use the technology than when we do not. The aim of our strategy is to use technology to make peer instruction efficient and effective in order to increase student learning, critical thinking, and communication in the teaching laboratory.

Our technique requires lab groups to answer questions regarding their predictions, observations, and interpretations of data while performing experiments. We then direct each group to discuss its answer with another group, which responded differently. This chapter describes how we use technology to monitor student responses and use these responses to group the students for brief periods of collaborative discussion. We have used this strategy in laboratory sections of introductory physics courses. It should be easy to adapt to other courses with laboratory sections, since implementation of this strategy requires only minor changes in laboratory procedure. Currently, we are working with both biology and chemistry professors to implement the strategy in these labs.

DESCRIPTION

Without radically modifying a given laboratory exercise, questions are added to focus student attention on the underlying physical concepts. These questions usually are one of three types. Students respond to (i) predictive questions—which type of image will the curved mirror produce; (ii) observational questions—what happened when you tried the experiment; and (iii) explanatory questions—why can't you produce an image on a screen with a convex mirror.

We embed these questions in the laboratory instructions. We try to insert these at key places where students are confused about what will happen or may have alternative concepts to explain the results.

Free-response questions may be used, but we usually select multiple-choice questions that include "none of the above" as one choice. Multiple-choice questions have several advantages: They force the group to choose one answer and to be prepared to defend that answer. They simplify record keeping. The students do not have to spend much time debating the wording of their answer and instead can concentrate on the concepts and the logical reasoning needed to support an answer. Including the option "none of the above" eliminates some of the disadvantages of using multiple choice format and leads to a modification of the choices to include more of the answers that students might give.

As each student group responds to these questions, their responses are sent to the instructor. Lab groups send the response via a networked computer. We use Beyond Question, an application that is similar to Classtalk, with networked computers to monitor student progress.[1] With Beyond Question, student responses immediately appear on the instructor's computer for easy assessment of the current classroom understanding of the given topic. Polling students using this technology has been successfully used to monitor student learning in large classroom lecture sections of introductory physics.[2] We have extended this into the physics laboratory. Responding via computers provides an organized record that can be saved electronically and increases the number of students that one instructor can monitor. The instructor notes the response of each group without making a comment and uses the responses to partner lab groups together.

Our strategy is for the instructor to watch for lab groups that give different answers to the questions. Then, the instructor asks two groups that have a conflict in answers to discuss these answers. These discussions are usually brief, but introduce a group to a different perspective, which the group either rejects or accepts. The groups do not have to reach consensus, but usually do. After the discussion, each group continues with the lab experiment. At the next question, the instructor may ask the students to have another group discussion, this time usually with a different group. This discussion procedure adds little to the time students spend in lab. In fact, the judicious pairing of a fast group with one that is having trouble often helps the latter to catch up. Also, the faster group, having explained their thoughts to the other group, results in understanding the concepts better. In general, students like to receive aid from and give aid to their peers.

Although all answers are reported, the instructor determines how frequently to pair student groups for discussion. Even with infrequent pairing, the reporting of answers keeps students on-task and increases student learning. This technique can be used with the lab instructions that are submitted to the student in written (paper and pencil) form or with instructions that are submitted electronically as a document or part of a Web page on a computer.

ADVANTAGES

There are a number of ways student learning improves. Students are encouraged to spend more time thinking about the experiment. They stay on task and make better observations because they have to report both to the instructor and to peers. In traditional labs, students who spend the most time trying to understand the experiment often stay in lab the longest. Most students' goal is to finish in the lab as quickly as possible. If students know they will have to defend an answer to another lab group, the dynamics change. They spend the time necessary to be able to defend their answer. Further, the students, in explaining to peers, are learning by teaching. Discussions with other lab groups also introduce a fresh perspective on a problem. Often in lab groups, the first person to give a reason becomes the authority. Without interaction between groups with other ideas, the first explanation can persist for the entire lab, even if it is incorrect.

Eric Mazur, who has developed a method for teaching physics based on peer instruction, found that when students are involved in peer instruction, students with scientifically accepted views generally are convincing to those with other views.[2] Mazur shows that peer instruction can lead to significant improvement in students' conceptual understanding. While Mazur uses peer instruction in the classroom with lectures, we have adapted it and developed it as we incorporate group-to-group discussion in the laboratory.

Along with employing peer instruction in the laboratory, our technique uses questions in the laboratory that are predictive, observatory, and explanatory. Research on student learning indicates that a cycle of prediction, observation, and then explanation promotes student learning.[3] As students make predictions, they have more ownership in the outcome. Sokoloff and Thorton, developers of the

Interactive Lecture Demonstrations pedagogical strategy, found that students who make and defend predictions understand concepts better than those who do not.[4]

Furthermore, this procedure moves laboratory exercises closer to the way scientific research is conducted. For example, if students disagree about experimental observations, then nature is the final authority. We find that students, on their own, repeat portions of an experiment and make new measurements to remove discrepancies and reconcile differences. Discussing possible interpretations of the data is also part of the scientific research experience.

Questioning each other and defending explanations encourages students to think critically about explanations. Students must understand and recognize logical inconsistencies and contradictions in their reasoning and the reasoning of the other group. These critical thinking skills are essential to enable students to evaluate their own and others' explanations. In order to make predictions about experiments, students apply abstract concepts to concrete observations. In order to interpret data and explain their observations, students develop abstract concepts from concrete observations. Our method encourages students to develop abstract reasoning and apply abstract concepts.

Promoting discussion across the laboratory will also improve communication skills. Students must not only explain ideas clearly to each other but also explain succinctly ideas to peers who were not part of the background discussion. They also must listen carefully to the ideas of someone else, a skill often neglected in education. Students practice explaining ideas on tests, in homework, and in response to questions posed in class, but they often are not required to listen to the ideas of their peers. Both explaining and listening are crucial for good communication skills and are a major goal of education.

RELATION TO CURRENT ACTIVE-LEARNING STRATEGIES DEVELOPED FOR PHYSICS

Current research in physics education focuses on developing techniques that actively engage students in their own learning. Our method is easy to adapt to a variety of these teaching strategies. Since students only report answers to the instructor and then discuss them with peers, the instructor is the facilitator, not the class expert, consistent with many strategies. As a facilitator, the instructor knows when it is appropriate to have the entire class discuss a particularly troubling point.

Our technique complements *Workshop Physics, Real Time Physics, and Tools for Scientific Thinking*,[5] all of which employ microprocessor-based workshop environments. To varying degrees, these methods use laboratory-based activities to guide students in their development of physical concepts and laws. For example, using computer-interfaced equipment, students measure their lab partners' movements as a function of time to understand that velocity is the change in position in a particular time. They compare this with a change in velocity, called an acceleration, which they also measure. Then, they measure the forces involved from which they develop the equation $F=ma$.

In these laboratory exercises, monitoring classroom progress on networked computers is a natural use of an available resource. (Monitoring student responses also helps the instructor quickly spot equipment that is malfunctioning.) The computer display of responses quickly directs the instructor to the group who would most benefit from further attention. Creating discussion groups lets students benefit from the comparison of predictions and verifications of observations as described above.

Lillian McDermott et al. have developed a set of tutorials based on common misconceptions that students have in physics classes, even after they have been introduced to material in the traditional lecture.[6] These worksheets force students to confront their misconceptions and move beyond them. For example, a tutorial on force diagrams asks students to draw the forces acting on a block and then discusses what to include in the diagrams (i.e., do not include forces that the block exerts on something else, a common mistake students make.) In these tutorials, students answer questions on worksheets aided by teaching assistants "who serve as facilitators, asking leading questions in a semi-Socratic dialog to help the students work through difficulties in their own thinking."[7] Our method can help teaching assistants be more aware of the concepts students are developing. By pairing student groups, the students can teach each other, relieving some of the burden from the teaching assistants.

Our method also seamlessly fits into using technology to teach physics based on a constructivist pedagogy in constructing physics understanding in a computer-supported learning environment (CPU project).[8] In this project, students follow computer-based, hands-on activities and construct their own understanding of physical concepts with very little direction from the instructor (traditional lectures do not exist). In small lab groups and in large class discussions, the students learn from each other and develop robust explanations of experimental observations. Our strategy can help the instructor monitor student progress and aid in the crucial cross-classroom discussion as students construct explanations.

Most lab instructors may be hesitant, reluctant, or unable to incorporate some of the more drastic modifications of lab mentioned above (*Real Time Physics*, MBLs [Microcomputer-Based Laboratory], etc.). However, instructors will find that predictive, observatory, and explanatory questions are often already embedded in traditional laboratory materials. Thus, our procedure can be incorporated into traditional labs very easily, allowing instructors to take advantage of many benefits of current physics education innovations without a drastic revision of the structure of laboratory and lecture sections. It transfers into the laboratory features of *Peer Instruction*[2], *Interactive Lecture-Demonstrations*[4], and other innovations, which have impressive data supporting their effectiveness.

PRELIMINARY RESULTS

To determine if this method indeed enhances student learning, we enlisted the help of two of our colleagues as their students did a physics laboratory experiment in optics. Both classes (multiple sections of the same course) used the same set of laboratory instructions, and for both classes these instructions included the embedded multiple-choice questions that we had prepared. One class used our strategy. The students in this class recorded their answers on the computer, which appeared on the instructor's computer and then, based on directions from the instructor, different groups discussed their answers with other lab groups. In the other class, the lab groups had very little interaction with each other. The students answered the same questions but did not give their answers to the instructor until they turned in the lab. They also did not discuss their answers with other groups.

The laboratory we used was a traditional physics lab. It was a geometric optics lab that included finding the focal length of curved mirrors and lenses, and image magnification. The lab essentially duplicated "Experiment 41: Lenses and Mirrors" in *Physics Laboratory Experiments*, a standard physics laboratory manual.[9] Thus, in this testing of our method we took a traditional lab, embedded questions in the lab instructions, and gave the instructions to two groups. (See Appendix A.) In one lab section, groups discussed their answers with other groups after sending the answers to the instructor, while in the other lab section, the groups did not.

In both labs, the students had not been introduced to the material in class, making the lab their introduction to this specific material on optics. Furthermore, in both classes, the instructor did not give an introductory explanatory lecture, but explained that the material students needed was included in the lab instructions themselves. Therefore, the instructors' explanations in both classes were primarily limited to answering procedural questions about the lab exercises themselves.

To measure how much students learned in the laboratory, we developed a 10 question test to administer to the class both prior to and following the laboratory. (See Appendix B.) We chose to use pre-tests and post-tests so we could focus on student learning gains in the lab. This also removed any bias in the data that might come from students who had previously studied this material. Furthermore, since the instructors gave no instruction on the laboratory concepts, any improvement in student scores from the pre-lab test to the same test taken after the lab, was due to what the students learned from working on the lab and not on the differences between instructors and types of instruction.

The student populations in both sections were similar in composition: a mix of primarily biology and marine science students along with a few chemistry and physics majors. The final course average for students in Lab A was B and Lab B was B+. The biggest difference in the two classes was the size. Lab A, which incorporated our method, was a class of 21 students; Lab B, the control group, had only 12 students. Unfortunately, we cannot control the class sizes, which are determined primarily by student scheduling conflicts.

For both classes, the pre-test average was 3.2 correct out of the 10 questions. The post-test averages for the two classes was significantly different. Lab B improved to an average of 5.5 correct out of 10. Lab A, whose lab groups were paired for discussion, improved to an average of 7.1 correct out of 10. Perhaps a more useful measure, though, is the fractional increase in percent correct, defined as where S_i and S_f are the pre- and post-test scores in percent. The value of the gain ranges from one (when a student answers all correctly on the post-test that she/he missed on the pre-test) to zero (student shows no improvement from pre- to post-test) or even negative values (student misses more questions on post-test than pre-test).

For Lab A, the average gain (G) was 0.62 (standard deviation: 0.07), while for Lab B, G was 0.37 (standard deviation: 0.08). While both groups showed learning gains from the laboratory, Lab A showed significant learning gains in comparison with Lab B. We are aware that this is a small data sample; nevertheless, this preliminary result shows clearly that this technique does indeed increase student learning. Lazarowitz and Tamir[10] have indicated that it is very difficult to obtain statistics that demonstrate the effectiveness of laboratory work. Thus, we find these preliminary results quite striking.

In comparison, note that for traditionally taught classes, the overall gain in conceptual understanding for the course (not for one particular topic as we tested) is $G \doteq 0.25$, while more interactive courses have gain values in the range of 0.36 to 0.68.[11] So the values of G we are seeing when our method is used are comparable to learning gains in interactive courses as expected for laboratory work.

This data shows that our technique can improve the laboratory experience in a variety of ways to increase student learning. Although harder to measure directly, our own experiences indicate that this improves the critical thinking and communication skills of the students as well. For example, a class culture of discussion can arise from this technique. In Lab B, because the instructor had used our technique in two previous labs, a culture of discussions between lab groups had already developed. The instructor rarely had to pair groups for discussion. The groups did it themselves and the instructor found them engaged in meaningful discussions that went to the heart of the challenging conceptual issues of the laboratory material. Therefore, even though students had to stop doing the experiment and talk with other groups (which is time consuming), the students voluntarily engaged in discussion about physics (and not the latest campus gossip!), improving both their communication and critical thinking skills along the way.

CONCLUSION

We have developed a technique using technology to monitor student responses in the laboratory and then pair lab groups for brief discussions. We found that the lab class that employed our strategy showed significant conceptual learning gains in comparison with another class engaged in the same laboratory, as measured by a set of pre- and post-tests. Further, the response of students, the changed attitude in the laboratory, and our own observations from using this strategy since the spring semester of 1997 indicate that our procedure has significant positive impact. Having students respond to questions increases their involvement and provides the lab instructor with a window into each lab group. This can be used to improve the laboratory experience in a variety of ways to increase student learning, critical thinking, and communication skills.

Endnotes

1. *Beyond Question* is available from William F. Junkin, Erskine College, Due West, SC 29639; 864-379-8822. *Classtalk* is available from Better Education, Inc., 4824 George Washington Hwy., Yorktown, VA 23692; 757-898-4846.

2. E. Mazur, *Peer Instruction: A User's Manual*, (Prentice Hall, New Jersey, 1997).

3. K. Tobin, D. J. Tippins, and A. J. Gallard, "Research on Instructional Strategies for Teaching Science," *Handbook of Research on Science Teaching and Learning*, ed. D. L. Gabel, (Macmillan Publishing Company, New York, 1994).

4. D.R. Sokoloff and R.K. Thorton, "Using Interactive Lecture Demonstrations to Create an Active Learning Environment," *The Physics Teacher*, 35 (1997) 340-347.

5. P.W. Laws, *Workshop Physics*, (Vernier Software, Portland, Oregon, 1995); P.W. Laws, D.R. Sokoloff, R.K. Thornton, *Real Time Physics*, (Vernier Software, Portland, Oregon, 1995); and D.R. Sokoloff and R.K. Thornton, *Tools for Scientific Thinking*, (Vernier Software, Portland, Oregon 1992 and 1993).

6. L. C. McDermott, P.S. Shaffer, and the Physics Education Group, *Tutorials in Introductory Physics*, (Prentice Hall, New Jersey, 1998).

7. E. F. Redish, J. M. Saul, and R. N. Steinberg, "On the Effectiveness of Active-Engagement Microcomputer-Based Laboratories," *American Journal of Physics*, 65 (1997) 46.

8. Both authors are part of the CPU project, supported by the NSF Grant ESI-9454341 and administered out of the Center for Research in Math and Science Education, San Diego State University.

9. J. D. Wilson, *Physics Laboratory Experiments*, 5th ed. (Houghton Mifflin Company, New York, 1998).

10. R. Lazarowitz and P. Tamir, "Research on Using Laboratory Instruction in Science," *Handbook of Research on Science Teaching and Learning*, ed. D. L. Gabel, (Macmillan Publishing Company, New York, 1994).

11. Mazur, p. 46.

APPENDIX A: QUESTIONS EMBEDDED IN LAB

For a Concave Mirror

1. Predict: When the screen is adjusted so that the object makes a clear image on the screen, the distance from screen to mirror will be:

   A. less than f

   B. f

   C. greater than f

   D. depends on the size of the object

   E. there will be no clear image

   F. none of the above

2. According to your ray diagram, when the screen is adjusted so that the object makes a clear image on the screen, the distance from screen to mirror will be:

   A. less than f

   B. f

   C. greater than f

   D. depends on the size of the object

   E. there will be no clear image

   F. none of the above

3. Which of the following light rays (going from the tip of the object arrow to the mirror as depicted in Figure 28.1) will bounce off the mirror and return parallel to the central axis? The center of curvature and focal point of the mirror are both shown on the diagram.

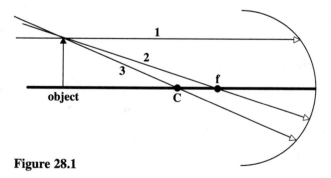

Figure 28.1

4. According to your ray diagram, the image will be:

    A. real, upright, and magnified

    B. virtual, upright, and magnified

    C. real, inverted, and magnified

    D. virtual, inverted, and magnified

    E. real, upright, and smaller

    F. virtual, upright, and smaller

    G. real, inverted, and smaller

    H. virtual, inverted, and smaller

For a Concave Mirror with Object at New Position

5. same as 4

6. After forming the image on the screen, move the screen and look at the reflection of the object in the mirror. The image that you see is:

    A. upright and magnified

    B. inverted and magnified

    C. upright and smaller

    D. inverted and smaller.

For a Concave Mirror with Object Inside the Focal Point

7. same as 4

8. same as 6

For a Convex Mirror

9. Predict: According to your ray diagram, when you look at the reflection of the object in the mirror, the image that you see will be:

    A. upright and magnified

    B. inverted and magnified

C. upright and smaller

D. inverted and smaller

For a Convex Lens

10. Predict: When the screen is adjusted so that the object makes a clear image on the screen, the distance from screen to lens will be:

    A. less than f

    B. f

    C. greater than f

    D. depends on the size of the object

    E. there will be no clear image

    F. none of the above

11. Which ray in your ray diagram will leave the lens parallel to the axis?

    A. the ray passing through the center of the lens

    B. the ray initially parallel to the axis

    C. the ray initially headed toward the focal point on the far side of the lens

    D. the ray initially headed toward the point on the axis that is a distance f from the lens on the near side

    E. none of these

For a Concave Lens

12. same as 4

APPENDIX B

Pre- and Post-Test

1. Light rays are coming into the mirror in the box. Two of these are drawn. Where is the image located? (See Figure 28.2.)

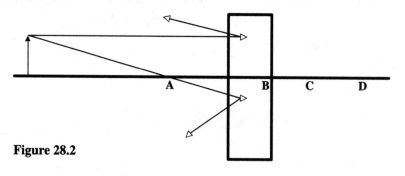

Figure 28.2

2. Light rays are coming into the lens. Two of these are drawn. Where is the focal point of the lens? (See Figure 28.3)

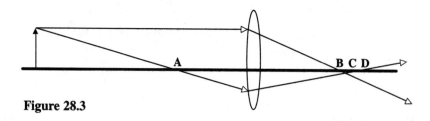

Figure 28.3

3. An object placed three meters in front of a lens with a focal length of two meters forms an image that is six meters on the other side of the lens. This image will be:

    A. bigger than the object

    B. the same size as the object

    C. smaller than the object

    D. the ratio of sizes depends on the size of the object

4. An object is placed slightly beyond the focal point *f* of the concave mirror. The ray diagram given below is drawn correctly. When a screen is placed so that the object is clearly focused on the screen, the screen will be located:

   A. at the focal point

   B. closer to the mirror than the focal point

   C. further from the mirror than the focal point

   D. the distance depends on the size of the object

   E. there will be no clear image no matter where the screen is placed

   F. none of the above (See Figure 28.4.)

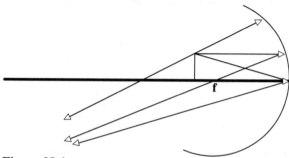

Figure 28.4

5. Which of the following is the correct ray diagram (and the correct location of the image) (See Figure 28.5 on page 376.):

6. Which one is the correct ray diagram for the convex mirror below? (See Figure 28.6 on page 377.)

7. Which of the following is the correct ray diagram? (See Figure 28.7 on page 378.)

8. A laser beam is hitting a shiny metal ball and bouncing off. Which way does the beam bounce off? The laser beam is the thick, dark line, and some possible choices for the way the light may bounce off are given with thin lines. Please circle the one that you feel is closest to the way it will actually bounce. The ball acts like a spherical mirror and both the center of the ball, C, and its focal point, f, are marked on the diagram. (See Figure 28.8 on page 378.)

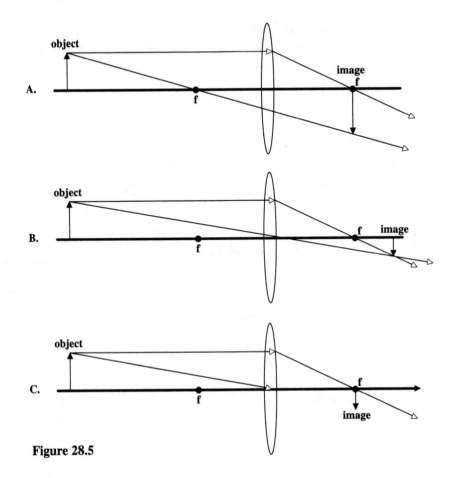

Figure 28.5

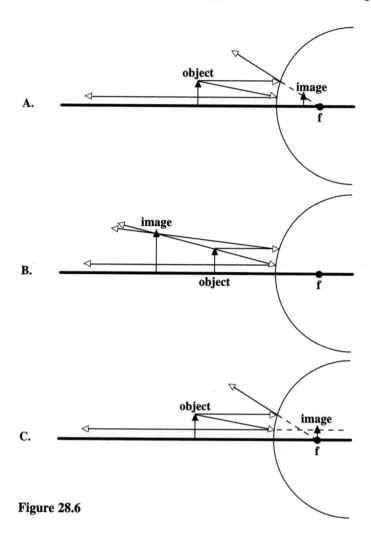

Figure 28.6

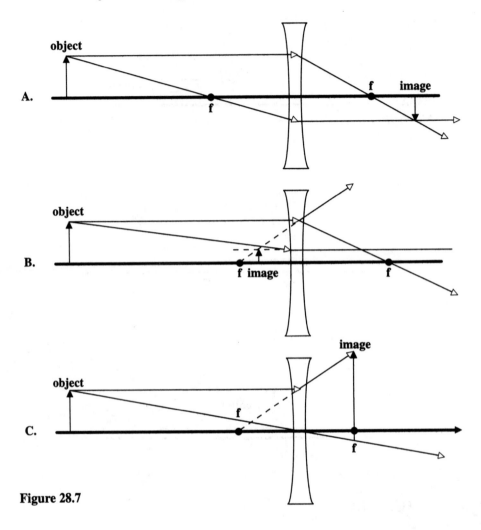

Figure 28.7

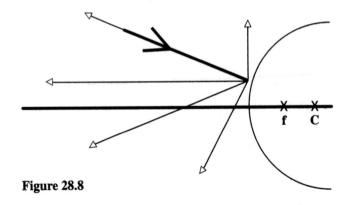

Figure 28.8

9. Which of the light rays (going from the tip of the object arrow to the mirror) will bounce off the mirror parallel to the horizontal axis? (Please circle the ray.) The center of curvature and focal point of the mirror are both shown on the diagram. (See Figure 28.9.)

10. For the diagram in Figure 28.9, where will the image of the object be located?
   A. to the right of the mirror
   B. at the focal point
   C. at the center of curvature
   D. there will be no clear image of the object
   E. none of the above

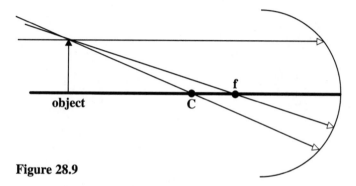

Figure 28.9

Index

U

Other Books of Interest from Information Today

Administrative Computing in Higher Education
Edited by Les Lloyd

With the expansion of campus-wide information systems and networks comes the advent of administrative computing—the use of networked computer systems by administrative personnel who share internal data as a function of their managerial duties. Issues discussed here include models of data sharing across systems, upgrading administrative software, selection, and expansion of computing systems, and other distributed computing topics. Contributors include a wide variety of educators and campus administrators who are directly involved in planning, building, and managing administrative computing systems in U.S. colleges and universities.

Hardbound • ISBN 1-57387-007-2 • $39.50

Technology and Teaching
Edited by Les Lloyd

Technology and Teaching presents an exhaustive selection of computer/multimedia applications case studies in various college courses, such as statistics, sociology, business management, writing, language translation, medical records programs, mathematics, music, and art. It even includes an outline for a World Wide Web workshop. The chapters provide information on hardware and software selection, whether ready-made or created by the college instructors themselves, as well as teaching strategies to work with the technology and to create software programs tailored to individual courses. There is something here for everyone in higher education.

Hardbound • ISBN 1-57387-014-5 • $42.50

Multimedia in Higher Education
By Helen Carlson and Dennis R. Falk

This book is a primer on the use of multimedia as an educational tool. It covers evaluation of teaching methods, selection of existing products, do-it-yourself product development, hardware and software needed, and more! Case studies of actual applications give readers concrete examples of the potential of multimedia computing systems in higher education. This book also offers definitions, examples of multimedia, guidelines for instructional design, and considerations for purchasing, repurposing, and developing videodiscs and other types of multimedia products.

Hardbound • ISBN 1-57387-002-1 • $42.50

Order online at www.infotoday.com

Information Today, Inc.

143 Old Marlton Pike • Medford • NJ 08055 • (609) 654-6266
E-mail: custserv@infotoday.com